ACTIVATED
SLUDGE
PROCESS

POLLUTION ENGINEERING AND TECHNOLOGY

A Series of Reference Books and Textbooks

EDITOR

PAUL N. CHEREMISINOFF

Associate Professor of Environmental Engineering
New Jersey Institute of Technology
Newark, New Jersey

1. Energy from Solid Wastes, *Paul N. Cheremisinoff and Angelo C. Morresi*

2. Air Pollution Control and Design Handbook (in two parts), *edited by Paul N. Cheremisinoff and Richard A. Young*

3. Wastewater Renovation and Reuse, *edited by Frank M. D'Itri* (out of print)

4. Water and Wastewater Treatment: Calculations for Chemical and Physical Processes, *Michael J. Humenick, Jr.*

5. Biofouling Control Procedures, *edited by Loren D. Jensen*

6. Managing the Heavy Metals on the Land, *G. W. Leeper*

7. Combustion and Incineration Processes: Applications in Environmental Engineering, *Walter R. Niessen*

8. Electrostatic Precipitation, *Sabert Oglesby, Jr. and Grady B. Nichols*

9. Benzene: Basic and Hazardous Properties, *Paul N. Cheremisinoff and Angelo C. Morresi*

10. Air Pollution Control Engineering: Basic Calculations for Particulate Collection, *William Licht*

11. Solid Waste Conversion to Energy: Current European and U.S. Practice, *Harvey Alter and J. J. Dunn, Jr.*

12. Biological Wastewater Treatment: Theory and Applications, *C. P. Leslie Grady, Jr. and Henry C. Lim*

13. Chemicals in the Environment: Distribution · Transport · Fate · Analysis, *W. Brock Neely*

14. Sludge Treatment, *edited by W. Wesley Eckenfelder, Jr. and Chakra J. Santhanam*

15. Wastewater Treatment and Disposal: Engineering and Ecology in Pollution, *S. J. Arceivala*

16. Carcinogens in Industry and the Environment, *edited by James M. Sontag*

17. Membrane Filtration: Applications, Techniques, and Problems, *edited by Bernard J. Dutka*

Additional Volumes in Preparation

ACTIVATED SLUDGE PROCESS

THEORY AND PRACTICE

Jerzy J. Ganczarczyk
Department of Civil Engineering
University of Toronto
Toronto, Ontario
Canada

MARCEL DEKKER, INC.   New York and Basel

Library of Congress Cataloging in Publication Data

Ganczarczyk, Jerzy
    Activated sludge process: theory and practice.

    (Pollution engineering and technology ; 23)
    Includes index.
      1. Sewage--Purification--Activated sludge process.
I. Title. II. Series.
TD756. G3   1983   628. 3'54   82-22127
ISBN 0-8247-1758-9

MARCEL DEKKER, INC.
270 Madison Avenue, New York, New York 10016

Current printing (last digit:
10 9 8 7 6 5 4 3 2

PRINTED IN THE UNITED STATES OF AMERICA

Preface

For more than six decades, activated sludge has been the basic biological wastewater treatment process around the world. However, since the relatively early monograph by Martin (Martin, A. J.: "The Activated Sludge Process," Macdonald and Evans, London, 1927) there have been no comprehensive works in the English language covering all aspects of the process applications in the form of a single publication.

The purpose of this book is to fill this gap, and to endeavor to present the updated information in a concise manner, while referring readers to more specific data in individual scientific and technical papers, and in various research reports. References were used from the literature up to the early part of 1981. The book was written basically for professionals working in the field of activated sludge treatment, such as consulting engineers, research scientists, and treatment plant supervisors. It can be also used as a specialized text for graduate studies. Care has been taken to keep a balance between the scientifics and practical aspects of the subject.

The writing of this book would not have been possible without the generous help of a number of my friends and professional colleagues. Numerous discussions which I have had with Dr. A. Pasveer (formerly of T.N.O., Delft, Holland), Professor W. von der Emde (Technical University of Vienna, Austria), Professor W. W. Eckenfelder (Vanderbilt University, Nashville, TN), and others, have contributed to the development of my approach to the subject. Parts of my draft manuscript have been read and commented on by (alphabetically): Professor B. J. Adams (University of Toronto), Mr. J. A. Barker (Greey Lightnin, Toronto), Dr. T. Chakrabarti (National Environmental Engineering Research Institute, Nagpur, India), Professor R. I. Dick (Cornell University), Professor M. F. Hamoda (University of Kuwait), Professor G. Henry (University of Toronto), Mr. P. J. Laughton (Anderson & Associates, Toronto), Mr. D. Prasad (University of Toronto), Professor W. O. Pipes (Drexel University, Philadelphia), and Dr. P. M. Sutton

(Dorr-Oliver, Stamford). In the description of flocculation phenomena, I
partially used a literature review prepared by my former student, Mr. D.
Carr. Special thanks are due to Mr. P. J. Laughton, Mr. D. Greey of
Greey Lightnin, Mr. Z. Mardirosian, formerly of Union Carbide, and
Mr. O. B. Burns, Jr. of Westvaco, for the use of their photographs. I am
also indebted to Mrs. Sandra Davey for her editorial suggestions, and to
Ms. Pat Scovil for her very fine and thorough typing of my manuscript.

    I dedicate this book to my wife who, in her subtle way, has encouraged
me to complete the work.

J. Ganczarczyk

Contents

Introduction

In general, the activated sludge process is a continuous or semicontinuous (fill-and-draw) aerobic method for biological wastewater treatment, including carbonaceous oxidation and nitrification. This process is based on the aeration of wastewater with flocculating biological growth, followed by separation of treated wastewater from this growth. Part of this growth is then wasted, and the remainder is returned to the system. Usually, the separation of the growth from the treated wastewater is performed by settling (gravity separation) but it may also be done by flotation and other methods (Fig. 1-1).

In microbiological terms, the above process can be classified as a continuous culture with sludge recycle. However, the principles developed for microbiological processes based on pure cultures growing on pure substrates are not directly pertinent to the activated sludge process in which mixed microbial cultures grow on mixed substrates at changing composition and concentration rates. Activated sludge treatment removes from the wastewater the dissolved and colloidal biodegradable organics as well as the unsettleable suspended solids and other constituents which can be sorbed on, or entrapped by, the activated sludge floc (Ganczarczyk and Obiaga, 1974; Matter-Müller et al., 1980, etc.). The mineral nutrients (phosphorus and nitrogen compounds) can also be partially removed, and some volatilization of purgeable organics may happen during activated sludge treatment (Pellizzari and Little, 1980; and Matter-Müller et al., 1980). One weakness of this method is that it does not remove the colour from many industrial effluents. In some cases, activated sludge treatment may even lead to increase the colour of the treated wastewater through the formation of highly-coloured intermediates resulting from the biological oxidation of certain organic compounds.

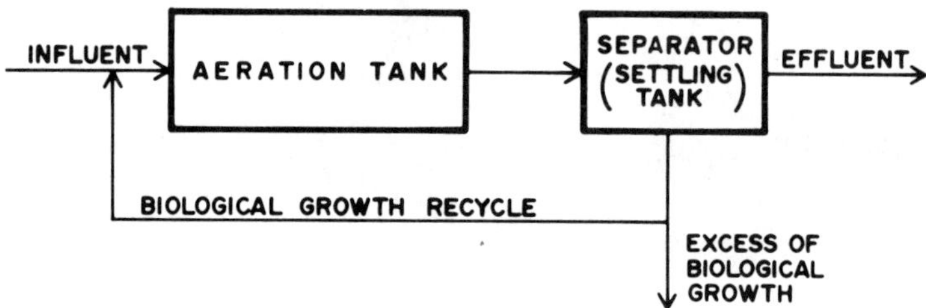

Figure 1-1. Basic concept of activated sludge treatment as a continuous process.

Use of the activated sludge process may be suggested for the treatment of small as well as large flows of wastewater, for domestic sewage, for mixed effluents and for industrial effluents only. At present, this process is, in North America, the most widely used method for the biological treatment of wastewater.

In the early stages of the development of the activated sludge process, fill-and-draw biological reactors (sequencing batch reactors) were used but were later abandoned in favour of continuous flow systems. Recently, this approach was re-evaluated by Irvine et al. (1977), and several of its advantages were shown, including equalization of flow and concentration, and total volume reduction. Although such systems were successfully used for the treatment of relatively small flows of various types of wastewater, their application has been very limited to date.

A biological alternative to the activated sludge process, which is in fact a suspended growth system, is any method which brings the wastewater into contact with a fixed biological film. Examples are the trickling filters and the rotating biological disks. The treatment of wastewater in fluidized biological beds (see Chapter IV p. 4.6) can be considered as an intermediate stage between the suspended growth systems of the bio-mass carrier and the fixed biological film covering the carrier particles. Similarly, the application of various types of activated sludge reactors (see Chapter V, p. 5.7), which are capable of retaining a certain amount of the bio-mass outside the recycle system, does not match precisely the basic definition of the process.

1.2 WASTEWATER AS A SUBSTRATE FOR THE PROCESS

There are substantial differences between the various types of wastewaters which can be treated effectively by the activated sludge process. These differences arise in:

1. concentration, composition and form of biodegradable and non-biodegradable organic carbon compounds;

2. presence and concentrations of mineral macro-nutrients (mainly
    N and P) and micro-nutrients;

3. presence and concentrations of substances which adversely affect
    biological treatment (such as toxic substances and inhibitory
    agents); and

4. concentration of total dissolved solids.

The organic compounds present in domestic sewage consist of biodegradable substances which are generally in a colloidal and suspended form.
In contrast, some industrial effluents may contain mostly soluble biodegradable organics, and they may represent only a fraction of the total content of organic substances in the wastewater. The organic substances
found in domestic sewage are mostly protein (40-60 percent), carbohydrate
(25-50 percent) and fats (10 percent). Industrial effluents may contain a
variety of organic substances at various relative concentrations. Furthermore, domestic sewage contains an excess of the mineral nutrients required
for biological treatment (see Chapter II p. 2.2) while some industrial effluents are deficient in some or all of these nutrients. Domestic sewage
usually does not contain any components toxic or inhibitory to biological
treatment. On the other hand, many industrial effluents do show such properties, and have to be specially "prepared" for activated sludge treatment
(see Chapter VIII, p. 8.1 and 8.2). The concentration of total dissolved
solids in domestic sewage reflects primarily their concentration in the
water supply, which is usually low (300 mg/L or less). Some industrial
effluents have characteristically high concentrations of dissolved solids
(e.g. over 1,000 mg/L).

The principal measurements of performance of the activated sludge
process, and the characteristics of its primary design parameters, are
usually based on an indirect determination of the removal of biodegradable
organic substances from the wastewater. This is usually based on the
standard five-day biochemical oxygen demand (BOD). BOD values (Standard
Methods, 1975) give an indication of all the different biodegradable organics
present in the wastewater and any changes in them, therefore, do not necessarily reflect the transformation of individual compounds.

BOD values of numerous biodegradable organic chemicals were determined, among others, by Heukelekian and Rand (1955), Pitter (1976), Bridie
et al (1979), etc. Table 1-1 presents a few examples of such BOD values as
compared with the respective COD (chemical oxygen demand) values and the
stoichiometrical oxygen demand for these compounds required for a complete
oxidation to water and carbon dioxide (Bridie et al, 1979). The ratio between
BOD value and COD value for a specific chemical compound is often considered as its biodegradability measure.

TABLE 1-1
BOD and COD Values of Some Selected Chemicals
(after Bridie et al., 1979)

Substance	S g/g	BOD		COD	
		g/g	% of S	g/g	% of S
Allyl alcohol	2.21	1.79	81	2.12	96
n-Butyl alcohol	2.59	1.71	66	2.49	95
Diethylene glycol	1.51	0.05	3	1.51	100
Benzene	3.08	2.18	71	2.15	70
Styrene	3.08	1.29	42	2.80	91
Toluene	3.13	2.15	69	2.52	81
o-Xylene	3.17	1.64	52	2.91	92
m-Xylene	3.17	2.53	80	2.62	83
p-Xylene	3.17	1.40	44	2.56	81
Acetone	2.21	1.85	84	1.92	87
Acrolein	2.00	0.00	0	1.72	86
Phenol	2.38	1.68	70	2.33	98
3,5-Xylenol	2.60	0.10	4	2.60	100

S = stoichiometric oxygen demand

1.3 HISTORICAL DEVELOPMENT

During the last few years of the 19th century and in the early part of the
20th century, many investigations were carried out on the treatment of
sewage by aeration. In general, the results of these studies were not very
promising. However, the findings from the experiments of Clark and
de Gage (1912) may be assumed to represent the invention of the activated
sludge process. They observed that, during the process of sewage aeration, sludge was produced and the presence of this sludge improved the
treatment effects. Unfortunately, the continuation of these studies led Clark
to the development of immersed aerated filters, which proved to be technically less important than the activated sludge process. In the meantime,
work in the proper direction was continued in Manchester, England by
Fowler, Ardern and Lockett. They determined some basic features of the

activated sludge process (Ardern and Lockett, 1914), and its continuous method was patented by Jones (1914).

The first activated sludge treatment plant was put into operation in Worcester (England) in April, 1916. This plant was designed for eight hours' aeration time and 25 percent sludge recycle. The aeration system applied was small-bubble compressed air diffusion with the ridge and furrow arrangement of air diffusers (see Chapter V). The first large activated sludge sewage treatment plants were constructed on the American continent from 1916 to 1926. They were treatment plants in Houston, Texas; Brampton, Ontario; Milwaukee, Wisconsin; Pasadena, California; and Chicago, Illinois (North Side). Later, treatment plants were constructed in Chicago, Illinois (Calumet); New York, N.Y. (Wards Island); Columbus, Ohio; Cleveland, Ohio; Baltimore, Maryland (Back River); New York, N.Y. (Tallmans Island); Chicago, Illinois (Southwest); and in other places (Greeley, 1945).

Contemporary development of the activated sludge process has been contributed to greatly by such researchers as W. Rudolfs, H. Heukelekian, C. Sawyer, A. Pasveer, K. Wuhrmann, W. W. Eckenfelder, W. von der Emde and others.

Specific milestones in the implementation of the process were:

1930s — development of step-loading concept;

1940s — development of partial treatment by so-called modified aeration;

1950s — development of bio-sorption and oxidation ditches; industrial wastewater treatment with mineral nutrients and without sewage addition;

1960s — development of efficient mechanical aerators;

1970s — development of commercial oxygen systems, deep-shaft system, and fluidized beds.

It is expected that the early 1980s will see the much needed development of control methods and automation in the activated sludge process.

REFERENCES

American Public Health Association, et al.: "Standard Methods for the Examination of Water and Wastewater", 14th Ed., 1975.

Ardern, E., and Lockett, W. T.: "Experiments on the Oxidation of Sewage Without the Aid of Filters", Jour. Soc. Chem. Ind., <u>33</u>, Nov. (1914).

Bridie, A. L., Wolff, C. J. M., and Winter, M.: "BOD and COD of Some Petrochemicals," Water Research (Brit.), 13, 627 (1979).

Clark, H. W., and de Gage, S. M.: "Aeration of Sewage as an Aid to Filtration," 44th Annual Report, Massachusetts State Board of Health, 1912.

Ganczarczyk, J., and Obiaga, T.: "Mechanism of Lignin Removal in Activated Sludge Treatment of Pulp Mill Effluents," Water Research (Brit.), 8, 857 (1974).

Greeley, S.: "The Development of the Activated Sludge Method of Sewage Treatment," Sewage Wks. Jour., 17, 1135 (1945).

Heukelekian, H., and Rand, M.: "Biochemical Oxygen Demand of Pure Organic Compounds," Sew. Ind. Wastes, 27, 1040 (1955).

Jones, W.: British Patent No. 22737 of November 19, 1914.

Irvine, R. L., Fox, T. P., and Richter, R. O.: "Investigation on Fill and Batch Periods of Sequencing Batch Biological Reactors," Water Research (Brit.), 11, 713 (1977).

Matter-Muller, C. et al.: "Non-Biological Elimination Mechanisms in a Biological Sewage Treatment Plant," paper presented at the Xth Conference of International Association on Water Pollution Research, Toronto, Canada, 1980.

Pellizzari, E. D., and Little, L.: "Collection and Analysis of Purgeable Organics Emitted from Wastewater Treatment Plants," EPA-600/2-80-017, March 1980.

Pitter, P.: "Determination of Biological Degradability of Organic Substances," Water Research (Brit.), 10, 231 (1976).

Principles of the Activated Sludge Process

2.1 CONCEPT AND STOICHIOMETRY OF
AEROBIC TREATMENT

The aerobic biological systems treating organic wastewater depend upon the physiology of heterotrophic organisms which, in the presence of oxygen, utilize the organic substances present in the wastewater as a carbon source for cell synthesis and as an energy source. Many aerobic heterotrophs, moreover, are capable of using oxygen from nitrite and nitrate (denitrification) to accomplish the substrate oxidation and cell synthesis in the absence of oxygen. Bacterial populations in aerobic conditions are capable also of converting ammonia nitrogen into nitrite and nitrate nitrogen. This process is called nitrification and autotrophic micro-organisms are mostly responsible for this, although some heterotrophs can also contribute to such transformations. In general, nitrification is a sequential process in which ammonia is oxidized to nitrite by organisms of <u>Nitrosomonas spp</u> and, subsequently, nitrite is oxidized to nitrate by organisms of <u>Nitrobacter spp</u>. Biological denitrification involves the conversion of nitrate and nitrite nitrogen primarily into nitrogen gas, but some nitrous oxide or nitric oxide may be also formed in this reaction.

The concept of aerobic treatment of organic wastewater may be presented conveniently in the form of simplified stoichiometric equations of substrate oxidation, cell synthesis, and endogenous respiration (Weston and Eckenfelder, 1955). These equations assume a general formula for organic substrates: $C_X H_Y O_Z$, and one empirical formula $(C_5H_7NO_2)_n$ suggested by Hoover and Porges (1952) for the composition of activated sludge micro-organisms. This latter formula yields a monomeric molecular weight of 113. Burkhead and Waddell (1970) found that the chemical composition of activated sludge was not constant and varied from C_4H_8ON to $C_8H_{15}O_5N$, depending on the stage of growth and the particular organic substrate used.

The equation of direct substrate oxidation may be written as:

$$C_X H_Y O_Z + (X - \tfrac{1}{4}y - \tfrac{1}{2}z)\, O_2 \xrightarrow{enzymes} xCO_2 + \tfrac{1}{2}y\, H_2O + \Delta H_I \tag{2-1}$$

Consequently, the equation of cell synthesis will be:

$$n\,(C_x H_y O_z) + nNH_3 + n\,(x + \tfrac{1}{4}y - \tfrac{1}{2}z - 5)\, O_2 \xrightarrow{enzymes} (C_5 H_7 O_2 N)_n + n\,(x - 5)\, CO_2 + \frac{n}{2}\,(y - 4)\, H_2O + \Delta H_{II}; \tag{2-2}$$

and of endogenous respiration:

$$(C_5 H_7 O_2 N)_n + 5n\, O_2 \xrightarrow{enzymes} 5nCO_2 + nH_2O - nNH_3 + \Delta H_{III} \tag{2-3}$$

The oxygen requirements for the substrate oxidation and cell synthesis are a function of the substrate characteristics, but according to this simplified stoichiometry, the calculated oxygen equivalence value for respiration of cell mass is 1.415 lb/lb (kg/kg). It has been shown, however, by Burkhead and McKenney (1969) that this value can vary at least from 1.07 to 1.77 lb/lb (kg/kg).

The first phase of the nitrification process may be presented as:

$$NH_4^+ + 1.5O_2 \rightarrow 2H^+ + 2H_2O + NO_2^- + \Delta H_{IV} \tag{2-4}$$

and the second one as:

$$NO_2^- + \tfrac{1}{2}O_2 \rightarrow NO_3^- + \Delta H_V \tag{2-5}$$

The total nitrification reaction is:

$$NH_4^+ + 2O_2 \rightarrow NO_3^- + 2H^+ + H_2O \tag{2-6}$$

The denitrification phenomenon follow the reaction:

$$NO_3^- + 6H^+ + 5e^- \rightarrow 0.5N_2 + 3H_2O \tag{2-7}$$

In the process of nitrification, experimental yield values for Nitroso-monas and Nitrobacter vary in a range from 0.04 to 0.13 g of cell matter synthesized per gram of ammonia nitrogen oxidized to nitrite and in a range from 0.02 to 0.07 g cell matter synthesized per gram of nitrite nitrogen oxidized, respectively (Painter, 1970). These yields are very low in comparison with the yields of heterotrophic micro-organisms oxidizing organic carbon compounds, which usually are in the range of 0.5 g of cells per gram of BOD removed (see p. 2.8). This difference is reflected in the composition of activated sludge which, in an average case of sewage treatment, may contain 97% of heterotrophs and only 3% nitrifiers.

The production of energy or energy consumption in these reactions (ΔH values) will be discussed later (p. 2.4).

The above simple stoichiometric equations of aerobic oxidation have rather a limited application for the practical design of unit processes. Theoretically, however, they may lead to the determination of oxygen and nitrogen nutrient requirements (see p. 2.2) as well as of the biomass production (see p. 2.8).

It is thought that, when contacting of wastewater with the micro-organisms of activated sludge occurs in the presence of dissolved oxygen, the suspended and colloidal solids, and, to some extent, the soluble organic substances in the wastewater are adsorbed on the surface of activated sludge flocs. At the same time, intensive biological activity converts part of the wastewater organics into a reserve food inside the microbial cells of the sludge. These two phenomena combined (Fig. 2-1) are responsible for the initial rapid BOD removal in the activated sludge process. Additional wastewater organic matter is progressively removed by continued aeration, and its rate of removal depends upon the remaining BOD and the concentration of the activated sludge. Block Diagram (Fig. 2-2) shows the expected mechanisms of organic substrates transformation in the activated sludge process. The total load of the wastewater BOD "A" is reduced by the load "B" which is "retained" in the treatment system, while the remaining load "C" is still present in the treated effluent. Then, the load "B" is the substrate for the direct biochemical oxidation (load "D") after the Eq. 2-1, and for the biomass synthesis (load "E") after the Eq. 2-2. In turn, the synthesized biomass is the subject of endogenous respiration (load "F") after Eq. 2-3. The remaining biomass (load "G") represents the excess of biomass produced during the activated sludge treatment (net biological growth).

2.2 NUTRIENT REQUIREMENTS

For effective long term biological treatment of organic wastewaters which are deficient in mineral nutrients, their nitrogen and phosphorus requirements, at least, have to be satisfied. The proper evaluation of these requirements has been the subject of numerous studies, e.g., Helmers et al. (1952), Hattingh (1963), Haltrich (1969), etc., from which it can be generally concluded that the requirements differ for different wastewaters and for the different methods and degrees to which they are treated. However,

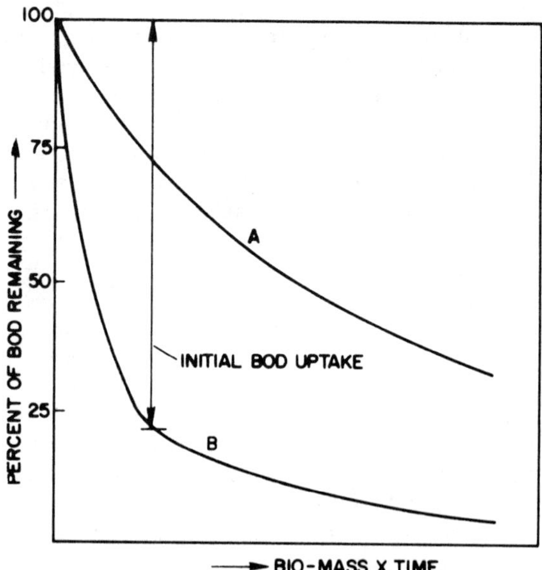

Figure 2-1. Single-phase (A) and two-phase (B) uptake of biodegradable organics by activated sludge micro-organisms.

they also depend on the hydraulic features of the treatment units and the mixing conditions of their contents, as these factors influence the rate of utilization of nutrients in the system and/or their rate of washout.

It is believed that the exogenous nitrogen requirements, expressed as a ratio of nitrogen to removable organic carbon, are related to the protein fraction of the biological growth, i.e. to the number of active bacteria in the system. This ratio, therefore, influences the effect of the treatment, the possibility of increasing the loadings and the stability of the process during flow and load variations. This may explain the decrease in the efficiency of the activated sludge process which is often observed as a consequence of nitrogen deficiency.

The performance of the activated sludge process in some nitrogen deficient systems was the subject of laboratory studies by Gaudy and Engelbrecht (1963), Ramarao et al. (1964), Komolrit et al. (1967), Gaudy et al. (1967), Goel and Gaudy (1968), and Gaudy et al. (1968). They showed that, for some period of time, the wastewater organic carbon can be removed effectively in the absence of exogenous nitrogen, and that this is connected with the increase of the carbohydrate and the decrease of the protein fractions of the activated sludge mass. Such systems can operate successfully with high contents of mixed liquor suspended solids and can be regenerated by the presence of nitrogen sources in exogenous or endogenous conditions.

This observation was confirmed in a full-scale experiment described by Ganczarczyk (1971). In an activated sludge treatment plant treating nitrogen deficient pulp mill waste, the supply of supplementary nitrogen was interrupted for a period of 25 days. This interruption in nutrient supply had very little influence on the treatment efficiency, but, subsequently, when the nutrient addition was re-commenced, some activated sludge was lost in the effluent and the sludge volume index increased. In this system, nitrification was completely inhibited and the only forms of nitrogen present in the system were ammonia and organic nitrogen. The concentration of ammonia in the final effluent was practically constant and independent of the addition of supplementary nitrogen. The explanation of this was that ammonia released from cellular materials occurs in proportion to the rate of endogenous respiration of the activated sludge.

Nitrogen and phosphorus are very important nutrients because the former is directly taking part in the biosynthesis, and the latter in the energy transfer. Other mineral nutrients such as Mg, K, Ca, Fe, Mn, Cu and Co are needed only in very small amounts (so called "micro-nutrients") and the mechanism of their influence is still not quite clear. It is often assumed that the concentration of trace elements present in wastewater is sufficient for good treatment. This assumption may not be valid in some circumstances because the types and concentrations of trace elements in wastewater vary with location, time, and possible wastewater pre-treatment. Trace elements may also be removed from the wastewater by various means, including adsorption and hydrogen sulfide precipitation. For these reasons, trace element deficiencies may occur, and treatment problems - including the occurrence of filamentous organisms - may result (Wood and Tchobanoglous, 1975).

The nitrogen and phosphorus nutrient demands for wastewater treatment were originally determined as optimal and minimal ratios to BOD

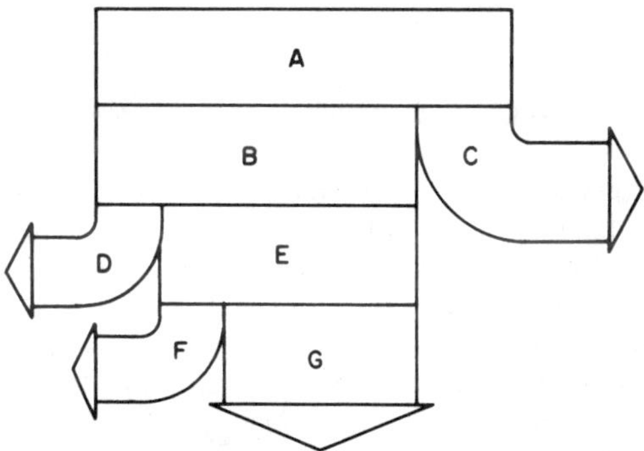

Figure 2-2. Block diagram of organic matter transformation in the activated sludge process.

loads (Sawyer, 1956). The commonly referred values for "optimal treatment" are:

$$BOD \; : \; N \; = \; 17.1 \text{ and} \tag{2-8}$$

$$BOD \; : \; P \; = \; 90:1 \tag{2-9}$$

The "minimal ratios" are accordingly:

$$BOD \; : \; N \; = \; 32:1 \tag{2-10}$$

$$BOD \; : \; P \; = \; 150:1 \tag{2-11}$$

A more universal determination of the demands of these nutrients can
be obtained on the basis of the balance of nutrients in the treatment system:

$$N \; = \; f'. \Delta V \; + \; c'.Q \text{ and} \tag{2-12}$$

$$P \; = \; f''. \Delta V \; + \; c''.Q \tag{2-13}$$

where: N and P are nitrogen and phosphorus nutrient requirements as nitrogen and phosphorus mass units per unit of time; ΔV is the net biological
growth in mass units per unit of time; f' and f'' are average fractions of nitrogen and phosphorus in the biological growth (e.g. from 0.07 to 0.12 for
nitrogen, and from 0.012 to 0.023 for phosphorus); c' and c'' are concentrations of nitrogen and phosphorus in final effluents, and Q is the flow of
wastewater.

In primary treated domestic sewage, the C:N ratio is usually below
2 - 2.5. However, the practical requirements of biosynthesis call for a
ratio of 5 - 6. Therefore, sewage can be a source of nitrogen for biological
treatment of nitrogen deficient industrial effluents. Other nutrients may
also be conveniently supplied by the addition of domestic sewage.

2.3 BIOLOGY OF ACTIVATED SLUDGE

Activated sludge represents a complicated mixture of viruses, bacteria,
protozoa and other organisms, found either singly or clumped together,
often enmeshed in a fabric of organic debris, dead cells and other waste
products. The response of this system to the environment is often difficult
to predict as individual organisms have varying sensitivities to nutrient
conditions, waste composition, and other stresses imposed upon the system
(e.g. inorganic salt concentrations, turbulence, pH, temperature and the
presence of competing micro-organisms). As a great number of these influences are hardly understood theoretically and certainly are not controlled
in a meaningful way in any operating treatment unit, many of the character-

istics indigenous to biological waste treatment systems, especially those
related to floc formation, remain largely outside any direct control.

BACTERIA OF ACTIVATED SLUDGE

Although organisms present in activated sludge systems range from viruses
to multicellular organisms, the predominant and most active are heterotrophic and, to a much lesser extent, autotrophic bacteria, which are both
aggregated in the sludge flocs and dispersed in the liquor (Fig. 2-3). The
importance of other organisms is often discussed only from the point of
view of their relationship to bacteria. Bacteria are unicellular organisms,
ranging in size from 0.5 to 5μ, and characterized by a relatively simple
cell structure. Heterotrophic bacteria utilize organic material as a source
of both carbon and energy, while autotrophic bacteria generally depend on
the oxidation of mineral compounds for energy requirements and utilize
carbon dioxide as a carbon source (e.g. nitrifying bacteria). Some species
of bacteria may utilize a wide variety of different organic compounds. In
heterogenous substances, specific constituents may or may not be prefer-
entially utilized. On the other hand, some substances which cannot be
easily utilized as a sole source of carbon and energy may be metabolized in
combination with other compounds. Usually a large number of different bacterial species are present in activated sludge. It appears that some industrial wastewaters may act selectively and drastically limit the number of
species present in the activated sludge. The physical characteristics of the
treatment system can also affect activated sludge population diversity by a
washout of slow-growers, elimination of species more sensitive to turbulence, etc.

Bacteria in activated sludge reproduce by binary fission (dividing), with
the generation time ranging from less than 20 minutes to a few days. The
typical bacterial growth pattern may be presented as a Buchanan's curve
(Fig. 2-4) for a batchwise substrate application. The original number of
bacteria "A", after being supplied with a dosage of a metabolic substrate at
time "t_1", first grows more slowly than the bacteria generation time, until
it reaches the number "B" at time "t_2". This first phase of bacterial growth
is called "lag phase", and it is believed that time "$t_2 - t_1$" is needed to acclimate the bacteria to the new environmental conditions. At the time "t_2",
the logarithmic growth phase ("log-growth") starts. During the period till
"t_3", bacteria divide at a rate determined by their generation time (constant
percentage rate of growth represented by a straight line on a semi-log
graph) as their growth is limited neither by a metabolic substrate nor nutrients' availability. After reaching the number of "C" at time "t_3", bacteria show a declining rate of growth, due to some growth limiting factors
(e.g. exhaustion of the metabolic substrate) and reach a "stationary phase"
at time "t_4", in which the growth of new cells is equal to the death of old
cells. This phase lasts till the time "t_5", and the number of cells "D" is
equal to the number "E". Then, the phase of "increasing death" follows

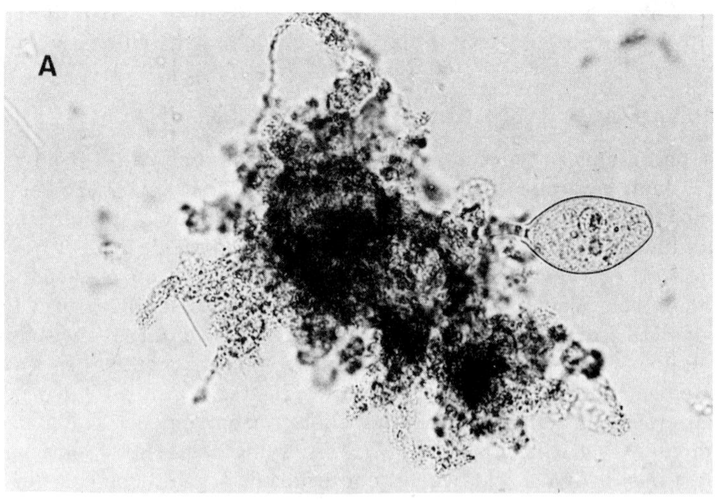

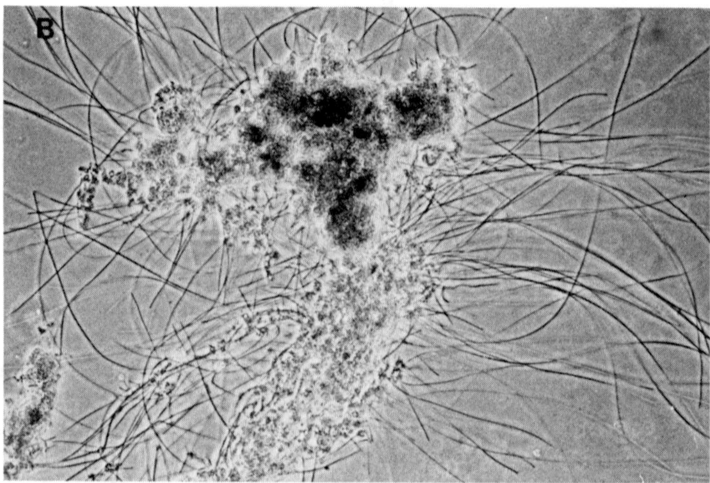

Figure 2-3. Microscopic photographs of activated sludge flocs at 390x enlargement (A = predominantly zoogleal floc; B = zoogleal floc enmeshed in a filamentous growth).

from the time "t_5" till the time "t_6". In this phase, less new cells are produced than old cells are lysed to produce metabolic substrate for the new cells. From time "t_6" until time "t_7", this phenomenon is characterized by a logarithmic rate ("log-death phase"), to stabilize in the period from "t_7" to "t_8" and eventually produce a number "H" of practically inactive cells. Basically, the number "H" is somehow larger than the initial number "A".

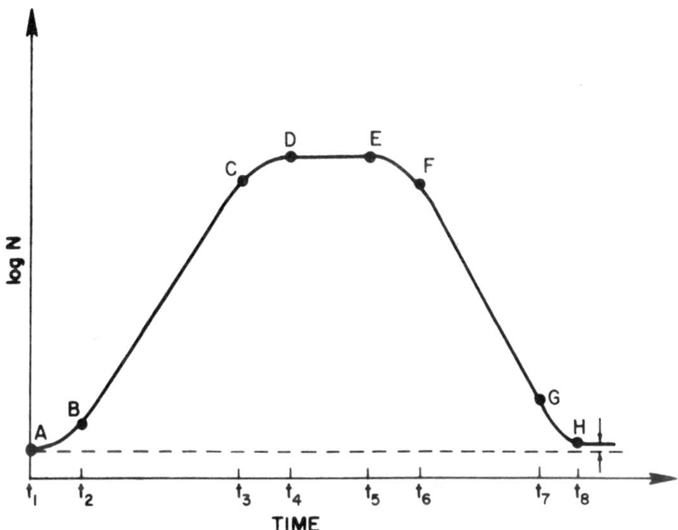

Figure 2-4. Buchanan's curve of typical bacterial growth pattern for a
batchwise substrate application.

In general, the activated sludge process involves aerobic bacteria, i.e.
bacteria utilizing oxygen for respiration. In practice, however, these bacteria may be exposed to anaerobiosis while inside the floc or when retained
in secondary clarifiers. Therefore, some can be expected to behave as
facultative micro-organisms, i.e. micro-organisms which are able to use
other electron acceptors when oxygen is absent.

It also appears that an intrinsic property of the heterogenous microbial
populations of activated sludge systems are the shifts in the species composition in response to several different environmental factors. The respective population dynamics may or may not result in changes in the physiologic-
al capacities of the treatment system.

The bacteriology of the activated sludge process was extensively reviewed by Van Gils (1964) and dealt with from a methodological point of view by
Lighhart and Oglesby (1969). Basic papers on this topic were also written
by Dias and Bhat (1964) and Prakasam and Dondero (1967). The flocculent
nature of activated sludge (see p. 2.4) makes it extremely difficult to enu-
merate the bacterial cell number in this material by plate count or other
similar procedures. However, the determination of DNA (deoxyribonucleic
acid) in biomass may be used as an indirect indicator of cell numbers
(Speece et al., 1973). It is believed that DNA exists only in the nucleus of
live and dead cells alike and that its amount per nucleus in haploid cells is
nearly constant.

FLOC-FORMING BACTERIA

Since the beginning of research into biological waste treatment, bacterial flocculation has received a great deal of attention. Early investigators (Butterfield, 1935; Heukelekian and Littman, 1939; and Wattie, 1943), concerned with the treatment of domestic sewage, noticed the prevalence of a floc-forming organism known as Zoogloea ramigera. This bacterium was found to possess a gelatinous matrix upon ageing and grew much better in aerated sterilized sewage than when cultured on nutrient broth supplemented with sugar. The addition of excessive amounts of carbohydrate caused dispersion of the floc due to the appearance of free-swimming cells, although attempts to induce bulking with glucose and starch failed (Butterfield, 1935).

Later studies by McKinney and Horwood (1952) and McKinney and Weichlein (1953) demonstrated that many other bacteria, isolated from sewage sludges, were capable of forming flocs as well. Over 30 bacterial strains derived from various activated sludge plants produced flocs when grown on sterile sewage or on a synthetic medium. McKinney was also able to show that flocculation was always associated with stationary or declining growth when a substantial part of the organic substrate had been exhausted (point "C" on Fig. 2-4). However, no attempt was made to determine the relative importance of each of the isolated organisms and it is unlikely that they hold as dominant a position as the Zoogloea bacteria. Similar studies by Dias and Bhat (1964) revealed a predominance of Zoogloea and Comamonas, although colonies of Aerobacter, Bacillus and Pseudomonas bacteria grew well on the vitamin and amino acid substrate used.

Some controversy over the floc-forming role of Zoogloea ramigera has arisen in recent years after several investigators uncovered strains of Zoogloea ramigera that neither produced a capsule nor a slime layer (Crabtree et al., 1966; Crabtree and McCoy, 1967; Angelbeck and Kirsch, 1969). This has led some authors to conclude that the formation of a Zoogloeal matrix by a bacterial strain is not synonymous with floc formation and that the two phenomena may not even be directly related (Friedman and Dugan, 1968). Later studies did reveal the presence of a micro-capsule, the problem being resolved when it was discovered that some Zoogloeal species produce a complex heteropolysaccharide polymer that does not take any of the known capsule strains (Friedman et al., 1968, and 1969). The capsule was revealed by layer electron scanning microscopy which also confirmed the presence of polymeric fibrils surrounding the Zoogloea bacteria.

The formation of a highly gelatinous, zoogloeal-like matrix which did not lead directly to flocculation could be attributed to the substrate used and the conditions under which the organism was grown. Considerable variation in cell size, shape and composition of the capsule and gelatinous material surrounding the cell has been observed by many researchers (Crabtree et al., 1966; Crabtree and McCoy, 1967; and Friedman et al., 1968). As many questions concerning the true form of zoogloeal growths go unanswered, the present state of taxonometric uncertainty and confusion surrounding the

genus Zoogloea is likely to exist for some years to come. Crabtree and
McCoy (1967), pointing to the vast array of conflicting data published on the
subject, concluded that Zoogloea ramigera, as reported in waste treatment
and other literature, is not a true species but only one of many kinds of
closely-related floc-forming bacteria. They gathered together twelve iso-
lates, all identified as Zoogloea ramigera and found that, of the twelve, five
Zoogloea ramigera, three were Zoogloea filipendula, two were Pseudomonas denitrificans and two were unidentified Pseudomonas.

Traditionally, the organism Zoogloea has been identified on the basis
of its flocculent growth habit (Butterfield, 1935; Wattie, 1943; Crabtree and
McCoy, 1967) and unique exocellular matrix (Breed et al., 1957; Dugan and
Lundgren, 1960; and Friedman and Dugan, 1968). But many other bacteria
have been shown capable of performing similar functions. In a normal activated sludge population, zoogleal bacteria are thought to be extremely pre-
valent and to contribute significantly to the development of a well flocculated
sludge mass. Visual observation of the cell and recognition of the characteristic capsule or matrix are usually cited as evidence of the presence of
zoogleal growths. However, with more and more previously unreported
variants being isolated from the sewage (Ganapati et al., 1967; Pasveer,
1969), these methods are increasingly being called into question by many
authors (Dugan and Lundgren, 1960; Friedman and Dugan, 1968; and Pike
and Curds, 1971).

FILAMENTOUS BACTERIA

The most commonly mentioned filamentous organism in connection with
activated sludge bulking is Sphaerotilus natans which is a bacterium dis-
playing some fungal properties. The development of this organism is usually induced by substrates rich in carbohydrates. There are also many
other filamentous organisms associated with sludge bulking, for example
fungus Geotrichum (Jones, 1965), sulphur bacteria: Beggiatoa and Thiothrix, etc. The identification of filamentous micro-organisms was discussed by Farquhar and Boyle (1971 and 1972). Until now, about 30 different organisms have been observed in activated sludge (Eikelboom, 1975).
Since so many different organisms may cause sludge bulking, it is very
difficult to find the right conditions to avoid their growth.

PROTOZOA OF ACTIVATED SLUDGE

The major taxonomic groups of invertebrates in activated sludge are the
flagellates, motile ciliates, stalked ciliates, amoebae, rotifers, and nema-
todes. In this system, many species of protozoa have been identified and
total numbers in the order of 50,000 cells per mL of mixed liquor are frequently recorded. It appears that when activated sludge systems undergo
changes (exposure to stress), ecological diversity declines but, with time,
the diversity may return again. There have been several attempts to assess

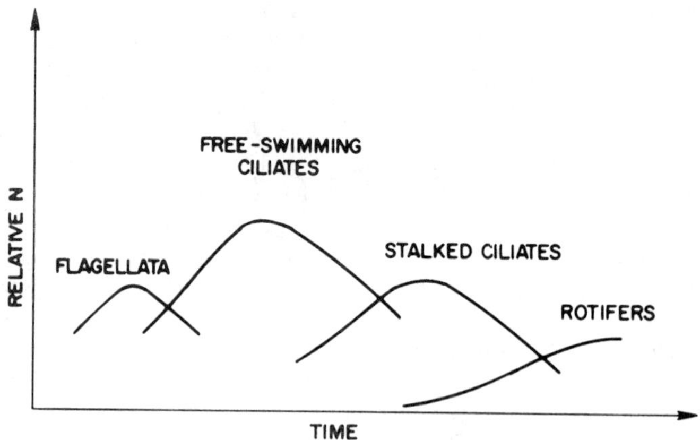

Figure 2-5. Succession of protozoa predomination in a start-up phase of activated sludge process (after McKinney and Gram, 1956).

the condition of sludge on the basis of the types and number of protozoa present. In general, a succession of predomination of protozoa during the start-up phase of the activated sludge process has been observed (Fig. 2-5). Initially, mostly flagellated and amoeboid protozoa are present, to be replaced by free-swimming ciliates, and then by stalked ciliates which are considered a sign of a well-operating process at a conventional loading rate. Rotifers are common only in over-aerated and under-loaded systems (McKinney and Gram, 1956).

Studies by Curds and Cockburn (1970) indicated that sludge loading affects the total number of species of ciliates present and the relative proportions of them in the sludge. Activated sludges maintained below loading values of 0.3 g BOD/g MLSS per day, in general supported the largest number of ciliate species, with an almost even distribution of species between different groups. The species structure of the protozoa population in activated sludge was found to be associated with the quality of effluent delivered. Certain protozoan species were more frequently found in plants delivering effluents within a particular effluent quality range than in plants delivering effluents outside this range. Using this data, it is possible to develop a method which enables the assessment of average effluent quality delivered from a knowledge of the protozoa species present in the sludge.

FUNGI OF ACTIVATED SLUDGE

Fungi are usually poorly represented in activated sludges but are common in activated sludges operating at lower pH values or for the treatment of certain wastewaters. Cooke and Pipes (1969) found that fungi were present in the largest numbers and species in autumn, declining with summer, spring and winter, in that order. In their free-living state, all fungi species used simple carbohydrates rather than complex substrates.

VISUAL EXAMINATION OF ACTIVATED SLUDGE

For many years, visual examination of mixed liquor has been used as an aid to operational control of the activated sludge process. Biological cultures developed on the wastewater from plastics manufacturing plants and containing marked concentrations of formaldehyde have a distinctive pink colour. Activated sludges from the treatment of coal carbonization effluents have a very dark colour (almost black), while sludges from sewage treatment usually have a light brown colour. Operators, who study the sludge from their own plant over a period of time, can develop a subjective concept of how the sludge should look when the process is performing properly. They usually use terms like "healthy", "fat", or "sick" to describe the appearance of the sludge. On this subjective basis of examination, each activated sludge process is unique and a feel for the appearance of sludge in one plant is not transferable to other plants.

2.4 FLOCCULATION OF MICRO-ORGANISMS

Although the factors which cause flocculent growths in mixed microbial cultures of activated sludge are not well understood, numerous theories have been advanced which partially explain such behaviour. The most common are the physical-chemical and biopolymer models. Much of the evidence which supports the latter hypothesis has been obtained in recent years from studies performed on pure culture bacteria. The observations made suggest that floc forming bacteria secrete an exocellular binding agent during the early declining or stationary phases of growth. Although its composition can vary greatly depending upon the micro-organism studied, the implications are that highly polymerized forms of extracellular polysaccharides and other polyelectrolyte-like compounds are responsible for bacterial agglutination.

CHEMICALLY INDUCED FLOCCULATION

Much work has been done on chemically induced flocculation of bacteria and colloidal dispersions through the addition of various synthetic and naturally occurring organic polymers, metallic ions and polyelectrolytes. Early research in this area concerned itself with a number of interparticle forces such as surface tension, specific ion adsorption and electrostatic charges (Buchanan, 1919). Flocculation was accomplished by the addition of simple electrolytes which reduced the repulsive surface potential of the stable suspension and allowed a closer state of aggregation when particles were brought together through a series of interparticle collisions as a result of Brownian movement. The inadequacy of this theory was soon evident when numerous other studies (Hodge and Metcalfe, 1958; and Busch and Stumm, 1968) demonstrated the fact that negatively charged polyelectrolytes and polymers quite often proved to be the best flocculents of negatively charged colloids. Likewise, many bacteria were observed to form stable suspensions very near their isoelectric point.

MCKINNEY'S CONCEPT OF BACTERIAL FLOCCULATION

McKinney (1952) has proposed a theory of floc formation in which capsulated bacteria are flocculated by direct chemical interactions between adjacent cells. Adsorption of cations from various salts in solution results in a reduced net surface charge which, under the influence of agitation, allows the cells to approach one another very closely. Because of the complexity of the surface structure of the bacteria and the presence of numerous enzymes, the possibilities for chemical exchanges are very large. Evidence suggests that peptide, salt and ester linkages are formed, assuming a possible reaction involving ionized carboxyl and amino groups on the bacterial surface.

Unfortunately, McKinney's emphasis on capsulated bacteria and later discoveries that non-capsulated bacteria demonstrate clear floc-forming abilities have led some authors to reject this approach to bio-flocculation (Pavoni et al., 1972). Later on, however, support was given by Tezuka (1969) to this theory. While working with a non-capsulated _Flavobacterium_, isolated in considerable quantities from a Tokyo sewage treatment plant (where it was considered to be dominant), it was noticed that the addition of a mixture of certain calcium and magnesium mineral salts flocculated the culture almost perfectly, regardless of what point in the growth or stationary phase the samples were taken. When the bacteria were supplied with mineral ions other than calcium and magnesium, no flocculent growth was observed. Suspending the flocculated bacteria in deionized distilled water caused a complete redispersion of the cells, while addition of small amounts of the necessary salts returned the cells to their former aggregated state. Negatively charged surfaces of adjacent cells were found to be bridged by ionic bonds intermediated by cations in accordance with the theory developed by McKinney. Paradoxically, when strains of _Zoogloea_ or _Pseudomonas_ were examined, the addition of a magnesium salt caused a redispersion of the suspension (Tezuka, 1967).

POLYELECTROLYTES AND FLOCCULATION

Following up on early successes with inorganic flocculents, a number of studies were performed to demonstrate the polyelectrolyte nature of chemically-induced bacterial coagulation (Tenney and Stumm, 1965; Busch and Stumm, 1968; and Dixon and Zielyk, 1969). Despite rapid progress

made in applying polyelectrolyte technology to wastewater treatment, no
clear and universally accepted model of flocculation developed from these
studies. For one thing, it appears that polyelectrolytes are highly selective
and sensitive to the material being coagulated. In this regard, the molecular weight and configuration of the polymer (i.e. long chain or cross-linked)
seems to be particularly important (Dixon et al., 1967). While in some
cases electrostatic repulsions between particles can be overcome with
polyelectrolytes, it is not universally true as numerous studies have
demonstrated (Singer et al., 1968; and Tenney et al., 1969). Other ano-
malies exist as well. For example, flocculation of a polymer system of
like charges will only be observed in the presence of an electrolyte,
usually a salt (Busch and Stumm, 1968; Stumm and Morgan, 1970). Varying
concentrations of electrolytes have been reported (Buchanan, 1919) but
even trace amounts seem to be adequate as long as a third ion is present.
Similarly, the effect of pH on different flocculents is quite complex and
poorly understood. Synthetic polyelectrolytes and trivalent metal cations
react quite strongly to changes in pH (Busch and Stumm, 1968; Tenney and
Stumm, 1965). Naturally occurring polymers also appear to be highly dependent on pH, although not all authors agree on this point (Busch and
Stumm, 1968; Pavoni et al., 1972). Hydrophilic colloids used as bacterial
flocculents have been found to be relatively insensitive to the pH of the medium (Hodge and Metcalfe, 1958). Unclear as the picture is, the complex
action of pH most likely results from its ability to simultaneously modify
both the nature of the polymer and the surface being destabilized (i.e. extent
of coiling of the polymer or the surface charge density on the colloid).

POLYMER BRIDGING MODEL

The polymer bridging model of coagulation developed by LaMer et al. (1957),
Healy and LaMer (1962), LaMer and Healy (1963) and subsequently applied
to cell aggregation by Tenney and Stumm (1965), Busch and Stumm (1968)
and Ries and Meyers (1968) is probably the most satisfactory approach,
and the best suited interpretation of floc formation at the present time.
According to this theory, electrostatic surface charge reduction to zero is
not required for flocculation to occur. It is postulated that adsorption of the
coagulant on the dispersed solids surface is first induced at specific
chemically-active sites where the free ends of polymer segments can come
into close contact with points on the uncoagulated colloid or cell. Physical
attachment of some of these segments leads to the formation of relatively
strong chemical or electro chemical bonds which, because of the short dis-
tances involved, are likely to overwhelm any local electrostatic repulsion
between particles having the same net charge. At very low polymer dosages,
many of the potential sites for attachment will remain unoccupied. As the
amount of polymer added is increased, these sites will become progressively
more saturated, leaving more and more of the polymer segments unattached.
With only a relatively small number of segments held rigidly on a fixed

number of chemically active surfaces, much of the polymer will remain in
solution as a series of extended segments. Under the influence of mixing,
these extended segments can make contact with each other or with sites
remaining vacant on other sites on the original surface. If the concentration of polymer is still further increased, bridging of polymers between
cells is observed and a three-dimensional matrix created. Under quiescent
conditions, this matrix will grow and eventually form a floc which will
settle out of solution, leaving behind a clear supernatant. Complete surface
saturation leads to a restabilization of the medium as a result of steric
(Heller and Pugh, 1960), electrostatic and physical interferences (Tenney
and Stumm, 1965; Tenney et al., 1969) between polymer segments. Mechanisms that have been considered for the attachment of polymers on predominantly negatively charged surfaces are hydrogen bonding, anion inter-
change with adsorbed anions (such as OH^-) or interchanges with cations on
or in the immediate vicinity of the colloidal surface (Busch and Stumm,
1968; Stumm and Morgan, 1970).

Experimental confirmation of the interdependence between the number
of surface sites available for adsorption of the coagulant and the degree of
coagulation achieved has been given by several previously mentioned authors and by Peter and Wurhmann (1971).

Although the general conclusions reached in the polymer bridging theory
(Fig. 2-6) have been verified experimentally in many fields, including waste
treatment, it still must be regarded as a first approximation to a better
understanding of bioflocculation. Many problems still remain to be solved.
Certain polymers, for example, seem to have an inhibiting rather than a
beneficial effect on the flocculation of some micro-organisms. Harris and
Mitchell (1972) worked with the bacteria <u>Leuconostoc</u> <u>mesenteroides</u> and
found that the production of dextran severely suppressed flocculation of the
cell over a wide range of pH values. Further experiments led the investigators to conclude that changes in the surface sorptive properties of the
dextran coated cells resulted in steric hindrances which prevented agglutination. Busch and Stumm (1968), in contrast to the above work, reported
that dextran was used to flocculate dispersed suspensions of <u>Escherichia</u>
<u>coli</u> and <u>Aerobacter</u> <u>aerogenes</u> and, therefore, they classed it as a floccu-
lant aid. Similarly, McGregor and Finn (1972) in a study involving the
yeast <u>Torula</u> <u>utilus</u> and bacteria <u>Pseudomonas</u> <u>flourescens</u>, found that
charge neutralization and polymer bridging failed to result in coagulation
of either organism in the presence of a cationic flocculant. In this case,
it was observed that biopolymers excreted by the cells competed for reac-
tive sites on segments of the cationic polymer and that the biopolymer-
polymer complex tended to increase the turbidity of the solution by sup-
pressing the formation of bridges between single cells.

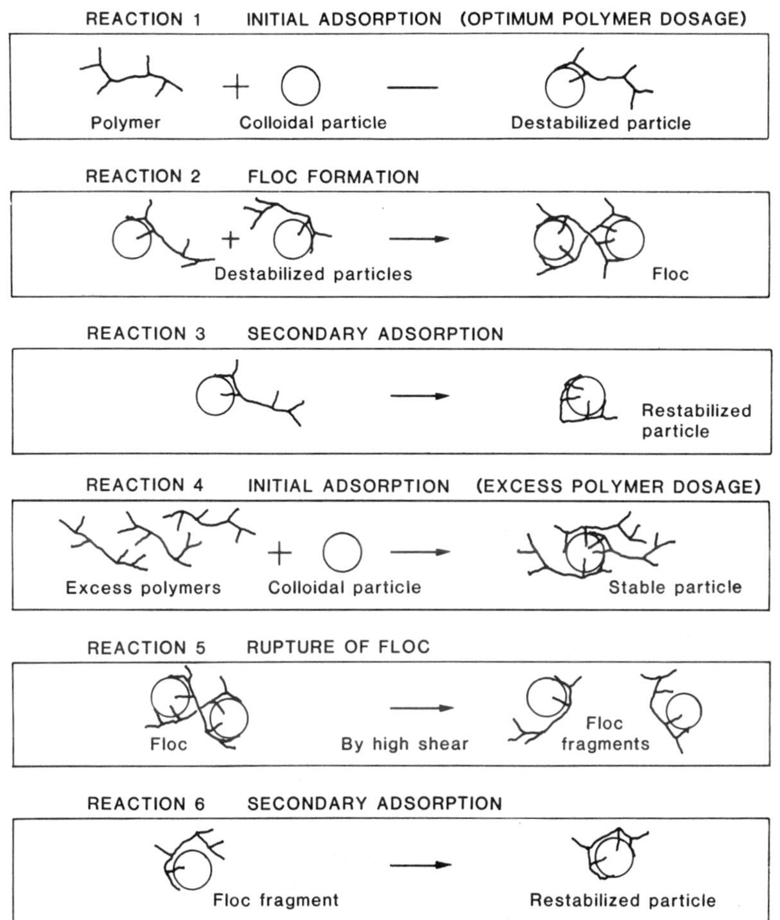

Figure 2-6. Polymer bridging model (after C. R. O'Melia).

BIOPOLYMERS IN ACTIVATED SLUDGE FLOCCULATION

Polymers originating from organisms found in sewage sludge have, until recently, received very little attention. In the past, most investigators were content to classify these materials only as volatiles (Walters et al., 1968) or as carbohydrates (Washington and Symons, 1962). Separation of microbial polymers from the bacterial cell wall has been accomplished physically by high speed centrifugation of the sludge sample (Busch and Stumm, 1968; and Pavoni et al., 1972) and by boiling (Wallen and Davis, 1972), or chemically by an acid or an alkaline extraction process (Davies, 1955; Dudman and Wilkinson, 1956; and Sandford and Conrad, 1966). Carr and Ganczarczyk (1974) reviewed these methods and evaluated them experimentally.

Currently, two very distinct and separate hypotheses on the origin of bacterial polymers have been advanced. On the one hand, it is claimed that these materials are derived basically from the waste metabolic products produced by the cells under conditions of nitrogen starvation and endogenous respiration. Fractional composition of this material reveals the presence of polysaccharide, protein, RNA, DNA and probably exocellular enzymes as well. In laboratory units, synthetic wastes, rich in sugars, have been found to produce the maximum amount of this material, although other carbon sources have also been used (Pavoni et al., 1972).

On the other hand, the relative abundance of nitrogen and the very low concentration of sugars found in most wastes have suggested to other investigators the possibility that floc formation is due to an entirely different cause. Accordingly, a second theory has been proposed in which carbon starvation is assumed to be the precursor of agglutination. According to this theory, it is probably humic acid materials which are polymerized to form the necessary long chain molecules that can act as bridging agents between bacteria. Infrared spectra analysis lends some support to this theory as it has been shown that characteristic humic-like materials do exist in sewage and activated sludges, while final effluents are almost entirely free of these compounds (Peter and Wuhrmann, 1971). In addition, the recent discoveries of important humic polymer groups in solids which chelate and adsorb many organic and inorganic compounds can be cited as further evidence of the potential importance of these materials in bioflocculation (Schnitzer, 1973).

Polysaccharides have received most attention among isolated exocellular microbial polymers. However, even the role of poly-β-hydroxybutyric acid (PHB) in promoting flocculation remains unclear. Evidence supporting the observation that PHB accumulations promote flocculation has been gathered by Dias and Bhat (1963), Crabtree et al., (1965), Crabtree and McCoy (1967), and Boyle et al., (1968). Other investigators (Rouf and Stokes, 1962; Friedman et al., 1968; Walters et al., 1968; Dean, 1969; etc.), while acknowledging that PHB will accumulate under certain conditions, assess its role primarily as a reserve storage produce which is utilized during endogenous respiration. Although varying amounts of PHB have been reported in a number of studies using both pure and mixed cultures, no direct correlation of PHB content to cell aggregation has been conclusively demonstrated (MacRae and Wilkinson, 1958; and Painter et al., 1968).

Many investigators (Duguild and Wilkinson, 1953; Dudman and Wilkinson, 1956; and Symons and McKinney, 1958) have shown experimentally that the amount of polysaccharide accumulated by the cell can be altered by

manipulating the ratio of carbon to nitrogen in the substrate. When nutrients
are present in excess, polysaccharides are almost entirely absent, except
upon rare occasions (Bukantz et al., 1941), particularly if the bacteria are
exhibiting log growth. As the external source of organic carbon is depleted,
some slight increase in this material is observed but it is not until the micro-
organisms have been subjected to extended periods of carbon and nitrogen
starvation that a rapid build-up of polysaccharide becomes apparent. Once
formed, the accumulated polysaccharide degrades quite slowly, leaving behind residual compounds which appear resistant to further degradation
(Symons and McKinney, 1958; and Washington and Symons, 1962).   Recently, Sato and Ose (1980) reported on the properties of viscous substances
extracted from activated sludge by sodium hydroxide solutions. These ex-
tracts contained protein, nucleic acids and carbohydrate.

2.5 ENZYMATIC REACTIONS, GROWTH FACTORS
AND ENERGY TRANSFER

In a biological treatment process, the decomposition of each waste water
constituent takes place through one or more pathways, which may involve a
number of intermediate compounds. Each step in this process includes a
specific enzymatic system, composed of one or more enzymes produced by
the micro-organisms active in the process. Enzymes are organic cata-
lysts. Chemically, they are proteins or proteins combined either with an
inorganic molecule or with a low molecular weight organic molecule. If the
production of any one of the required enzymes is inhibited, the entire chain
reaction may be blocked and the accumulation of an intermediate compound
may occur. The inhibitors of biological treatment do not necessarily have
to be present in the process substrate (wastewater). They can be produced
in the process itself by biotransformation and/or by the production of metabolites or intermediate compounds whose rates of decomposition are significantly slower than their rates of formation.

In the activated sludge process, there are two general types of enzymes:
extracellular and intracellular. Very often extracellular enzymes are active in hydrolysis of more complex organic compounds which are originally
unable to enter the cell wall. Intracellular enzymes are involved in the oxidation and synthesis reactions within the cell.

The enzymatic pathways required for decomposition of carbohydrates,
proteins and fats are numerous and no inhibition should be expected for biological oxidation of such common substrates. Conversely, oxidation of some
organic substances present in industrial effluents may include very selective
enzymatic mechanisms and, therefore, may be very easily inhibited by a
blockage of at least one of the reactions involved. It has proven possible,
however, to identify several compounds (growth factors), which effectively
by-pass such a blockage of the respective enzyme reaction when added in
small quantities to the process reactor, thus achieving an acceleration of the

biological decomposition. This is believed to be one of the reasons why
some industrial effluents are easier to treat in a mixture with other effluents. Catchpole and Cooper (1972) found that the addition of as little as
5 mg/L of p-aminobenzoic acid (PAB) or even glucose to unlimed coke-
oven effluents allowed a significant increase of the organic loading to the
aeration tanks used for treatment of this wastewater and a corresponding
reduction of the required hydraulic retention time. It was hypothesized that
the formation of pyruvic acid played a key role in the mechanism of the improved treatment process.

Various commercially available biocatalytic additives are sometimes
recommended to improve the operation of biological treatment of some industrial effluents, especially if there are some difficulties with the process
starting-up procedures. These additives usually contain a mixture of technical grade enzymes, dried bacterial solids, and by-products of bacterial
fermentation. The application of the additives, contrary to often overstated
claims of the suppliers, usually is not of much value and substantial costs
are often incurred. However, under special circumstances, it may make
sludge acclimatization easier and shorten the time taken to start the process.

Biochemical oxidations release energy and biochemical syntheses require energy. The main substance in the respective transfer of energy is
adenosine triphosphate (ATP). ATP is formed from adenosine diphosphate
(ADP) by adding inorganic phosphate via substrate level or oxidative phos-
phorylation mechanisms. Conversely, ADP and inorganic phosphate are
formed from ATP with the release of energy. The efficiency with which microbial cells can produce ATP and use it for synthesis is not fully known,
and probably varies from species to species.

Servizi and Bogan (1963 and 1964) postulated that cell yield or the synthesis reaction is proportional to the quantity of ATP formed per unit of
substrate. They further postulated that ATP production is proportional to
the free energy released through oxidation. Evidence as to the validity of
these proposals was obtained by analyzing the existing data of other authors
and from the results of their own studies. Following upon these ideas,
McCarthy (1965) found that the decrease in the free or useful energy of the
substrate times the efficiency of conversion of substrate energy to ATP energy is equal to the ATP energy required for synthesis and maintenance. He
proceeded to show that the ATP free energy for synthesis is the sum of two
basic energy steps. The first energy step, depending upon the original
energy state of the substrate, is to bring the cell carbon source to an intermediate level which is common to all organisms. The second energy step in
synthesis is to convert the intermediate into cellular material. The sum of
both steps gives the total energy for synthesis. Schroeder and Busch (1966)
have shown several limitations to the above approach.

As biological oxidation of any wastewater organics involves the removal
and transfer of electrones or hydride ion to a suitable hydrogen acceptor,
the activity of dehydrogenase enzymes may be useful as a measure of the
overall activity of the activated sludge process. Ford and Eckenfelder (1967)

and Jones and Prasad (1969) observed that both oxygen uptake rates and
dehydrogenase activities increased with increased sludge loadings for several treated wastewaters.

2.6 TRANSPORT AND DIFFUSION PHENOMENA

The stabilization of the organic substrate by activated sludge microorganisms occurs inside the bacterial cells. Therefore, the transport of substrates into the cell and that of metabolic products out of the cell is especially important. Large molecules of organics present in wastewater
have to undergo a preliminary degradation prior to permeation through the
cell membranes. Enzymes active in such transformations are synthesized
by bacteria and are either attached to the cell wall or released into mixed
liquor (extracellular enzymes). In some instances, simple diffusion of
metabolites through cell membranes does not explain the relatively high
rates of such transport which are observed. Therefore, a concept of so-
called "active transport" has been elaborated. Active transport means the
substrate transport against a concentration gradient across the cell membrane, and involves ATP as the source of energy along with a supportive
permease enzyme system.

The major transport phenomena involved in activated sludge treatment
(Swilley et al., 1964) comprise the interphase transport of:

1. gaseous oxygen and solid phase substrate into solution;

2. dissolved oxygen and substrate to the vicinity of a metabolizing
    organism;

3. dissolved oxygen and substrate into the cell; and

4. products of the biochemical reaction out of the cell.

Most of these transfer mechanisms are diffusional in nature and the values
of the respective diffusion coefficients may control the overall kinetics of
the process. Mixing and turbulence in the aeration tanks (see Chapter III,
p. 3.3) alter the liquid phase transport and decrease diffusion resistances.

The diffusion of oxygen to activated sludge flocs was the subject of
numerous studies (Pasveer, 1954; Wuhrman, 1963; Mueller et al., 1966;
Matson et al., 1972; and others). It is believed that the rate of oxygen
diffusion had a more pronounced influence on the process control than the
diffusion of other substrates. Baillod and Boyle (1970) studied intra-
particle diffusional resistances in the uptake of organic substrate by the
microbial floc. They established the existence of mass transfer limitation
in substrate removal by measuring glucose and oxygen uptake rates for
large activated sludge flocs and the same flocs after blending. Matson and
Characklis (1976) demonstrated also that the respective molecular diffusion

coefficients depended on sludge age values and carbon to nitrogen ratios of wastewater.

Activated sludge flocs may also remove some non-biodegradable wastewater constituents because of sorption or sorption related phenomena. This was demonstrated by Ganczarczyk and Obiaga (1974) for lignins present in pulp mill effluents (see Chapter VIII, p. 8.5), and by Neufeld and Hermann (1975) and Cheng et al., (1975) for heavy metals.

2.7 SUBSTRATE UTILIZATION

Biological uptake of biodegradable organic substances from wastewater in the activated sludge process is a complex phenomenon. Its mechanism depends on both the type of metabolic substrate and the process parameters, and may be explained in different ways. In addition to the growth rate of micro-organisms, other mechanisms like the substrate storage may considerably contribute to the effects observed. In the presence of multiple sources of organic carbon and energy, as in the case of most wastewater, activated sludge micro-organisms at loadings in the range of partial biological treatment may exhibit a preferential utilization of some substances and repression of the others (Gaudy, 1962 and 1963; Stumm-Zollinger, 1968; Ghosh et al., 1972). This fact is often ignored when total organic concentration is expressed in terms of BOD, COD or TOC. Some apparent fluctuations in the effectiveness of the process in removing organics from wastewater may also be a function of the fluctuating flocculation ability of the sludge as residual suspended solids can reflect on determinations of residual organics. Determinations on filtered effluents can help to avoid such uncertainty.

PROCESS KINETICS

The kinetics of the activated sludge process can be expressed by substrate utilization and biological growth relationships. These relationships are influenced by several parameters, which are not always fully recognized by specific approaches. This is well indicated by the diversity of existing kinetic theories which vary from zero-order to second-order reactions, and assume one-phase or two-phase phenomena.

The majority of the kinetics relationship have been developed in the studies of pure microbial cultures, using soluble and biodegradable substrates. Therefore, their practical application to wastewater treatment is quite limited. Further limitations arise with the requirements of producing low organic carbon effluents and possible complete separation of biomass from treated wastewater. These call for a food limited bacterial growth, and therefore, the reaction kinetics cannot reflect the logarithmic growth rates. It also has been found that continuous, completely-mixed systems performed at a higher unit rate of reaction than predicted by batch tests for the same substrate concentration. Moreover, the treatment results achieved (if expressed in residual organic carbon levels) have often been higher in continuous systems than in batch experiments. It also has to be recognized that the process turbulence (Chapter III, p. 3.6) and the process temperature (Chapter III, p. 3.7) increase the reaction rate, most likely by reducing diffusional resistances (p. 2.5).

One of the common approaches to activated sludge process kinetics (although not fully justified scientifically) is the application of the Monod equation (1949) for the evaluation of the rate of limiting substrate biological growth:

$$\mu = \hat{\mu} \frac{S_*}{K_S + S_*} \tag{2-14}$$

where: μ = growth rate of micro-organisms, day^{-1}; $\hat{\mu}$ = maximum growth rate of micro-organisms, day^{-1}; K_S = half velocity constant (substrate concentration in mg/L at half of the maximum growth rate): and S_* = growth limiting substrate concentration, mg/L

The above expression is similar to the Michealis-Menten equation for enzyme-catalized transformations.

When the substrate concentration S is low with respect to the value of K_S, the above equation may be simplified into a first-order equation as suggested by Garrett and Sawyer (1952):

$$\mu = \frac{\hat{\mu}}{K_S} S_* = KS_* \tag{2-15}$$

where: $K = \dfrac{\hat{\mu}}{K_S}$

Wuhrman (1957) and Tischler and Eckenfelder (1969) showed that the removal of specific substances in the activated sludge process usually was of zero order (linear), but, when considering the total substrate remaining at any times in terms of composite indicators like COD, BOD, TOC, the removal mechanism approximated first order kinetics. It was really a simulation of multiple zero order reactions. According to Tischler and Eckenfelder (1969) the batch removal of a mixture of organic substances by activated sludge can be approximated by the relationship:

$$\frac{S}{S_o} = e^{-kX_v t} \tag{2-16}$$

which for a continuous treatment system at constant food to micro-organism ratio becomes:

$$\frac{S_o - S}{X_v t} = kS \tag{2-17}$$

For concentrated wastewater (Adams et al., 1975) this relationship should be modified to the form:

$$\frac{S_o(S_o - S)}{X_v t} = KS \tag{2-18}$$

where: S_o = initial substrate concentration, mg/L; S = effluent substrate concentration, mg/L; t = hydraulic detention time, days; X_v = MLVSS, mg/L; k = specific substrate removal rate coefficient, L/mg day; and K = specific substrate removal rate coefficient L/day.

KINETICS OF SEPARATE STAGE NITRIFICATION

Nitrification rates for mixed cultures in one-stage activated sludge process treatment (combined carbon oxidation-nitrification systems) differ widely due to the differences in pH, temperature and composition of sewage. For example, according to Wild et al., (1971) an oxidation of 0.135 g NH_3/g MLSS per day was obtained at pH of 8.5 and at 20°C. Bishop et al. (1974) reported an oxidation of only 0.03 g N-NH_3/g MLVSS per day at 15.5°C and 0.11 g N-NH_3/g MLVSS per day at 27°C. These differences are due to different sludge ages (see Chapter III), and different influent C to N ratios which affect the contents of active nitrifiers in the mixed liquor volatile suspended solids.

Of course, in the absence of inhibitors, the respective values are much higher for the enriched cultures. Parker et al. (1975) even referred to a rate of 2 g N-NH_3/g MLVSS per day as a characteristic value. However, even in the separate stage nitrification, nitrifiers are only a relatively small fraction of the total bacterial mass (about 10 percent in many practical situations).

The maximum denitrification rates as observed for concentrated nitrate wastewater by Jewell and Cummings (1975) were 0.4 g N-NO_3^- rem./g MLSS per day for the suspended growth system and 0.5 g N-NO_3^- rem./g MLSS per

TABLE 2-1
Substrate Removal Rate Coefficients
(after: Eckenfelder, 1970)

Wastewater	k
Domestic	0.017-0.03
Refinery	0.074
Chemical & Petrochemical	0.29-0.018
Pharmaceutical	0.018

day for the attached growth system. The practical values found by Bishop et al. (1974) are much lower. They were equal to 0.03 g $N-NO_3^-$/g MLVSS per day at 15.5°C and 0.055 g $N-NO_3^-$/g MLVSS per day at 25°C.

RESIDUAL ORGANIC SUBSTANCES

Analysis of the organic matter in the filtered effluents from activated sludge treatment (DeWalle et al., 1975) indicated that significant differences existed among the effluents obtained from systems having different sludge ages (see Chapter III, p. 3.4). A decrease in the high molecular weight humic carbo-hydrate-like materials, with concomitant increase of the lower molecular weight fulvic-like materials was observed in the effluents from systems with increasing sludge ages.

Removal of residual organics in activated sludge treatment may be enhanced by additions of powdered activated carbon to the treatment system. This idea has been put into practice in a modification of the process called PACT Process (see Chapter IV, p. 4.11). A study performed by DeWalle and Chian (1974) revealed that activated carbon colum could retain from 70 to 80 percent of residual organic material present in activated sludge treated sewage. High molecular weight humic carbohydrate-like complexes and some amino and fatty acids were not removed. Treatment of the activated sludge effluent with a commercial cellulose acetate reverse osmosis membrane gave only 60 percent total organic carbon removal. Analysis of the retentate showed mostly the humic carbohydrate-like complex and fulvic acids. The combined use of activated carbon and reverse osmosis was not successful in the removal of all organics from the effluent.

The 0.5 to 3.5 mg/L of soluble organic nitrogen which is present in biologically treated effluents (Perkin and McCarty, 1974) consists of a refractory fraction and a slowly biodegradable fraction. The latter fraction represents from 50 to 70 percent of residual soluble organic nitrogen, and decomposes at ratios between 2 and 9 percent per day.

2.8 BIOMASS SYNTHESIS AND OXYGEN UTILIZATION

When the micro-organisms in the activated sludge process are brough
into contact with organic matter present in wastewater, a rapid increase
in biomass occurs, accompanied by a respective removal of the substrate
and an uptake of oxygen. In 1951 Heukelekian et al. proposed a general
way of activated sludge growth rate evaluation by adopting the formula

$$\Delta X_V = aS_r - bX_V \tag{2-19}$$

where: ΔX_V = accumulation rate of activated sludge in lb/day (or g/day) of
volatile suspended solids; S_r = BOD removed from wastewater in lb/
day (or g/day); X_V = mixed liquor volatile suspended solids (MLVSS
in lb (or g); a = synthesis rate in lb/lb BOD removed/day (or g/g BOD
removed/day); and b = MLVSS oxidation rate in lb/lb MLVSS/day (or
g/g MLVSS/day).

The above relationship can also be presented in the form of a differential
equation

$$\frac{dX_V}{dt} = a\,\frac{dS}{dt} - bX_V \tag{2-20}$$

which is identical to the formulation adopted later by Lawrence and McCarty
(1970). Eckenfelder (1965) presented experimentally determined coefficients
of sludge production and respiration for different types of wastewater (Table
2-2).

Activated sludge growth is apparently a function of particulate matter
retention, substrate storage by micro-organisms and their replications.
Studies of Speece et al. (1973) indicated that the phenomenon of substrate
storage was a function of sludge loading. Sludges at lower loading rates
showed an extended time lag before cell replication occurred, while higher
loading rates maintained the bacteria in a more active state. Practical
examples of the application of the process show sludge production values
from as high as 1 lb (gram) per lb (gram) of BOD removed to as little as
0.1 lb (gram) per lb (gram) of BOD removed. Clarification of wastewater
prior to activated sludge treatment substantially decreases the excess of
activated sludge and thus simplifies the problems of sludge treatment and
disposal. Figure 2-7 shows the usually observed range of sludge production
as a function of food to micro-organisms ratio in activated sludge treatment
of clarified sewage. Under cold climate conditions somewhat higher sludge
production may be expected. A more refined methodology for sludge
production analysis was recently offered by Benedict et al. (1979).

TABLE 2-2

Coefficients* of Sludge Production and Oxygen Requirements
(after: Eckenfelder, 1970)

Wastewater	Synthesis Rate "a" in $\#/\#BOD_r$	Oxygen Requirement "a'" in $\#O_2/\#BOD_r$	Rate of Endogenous Metabolism "b" in $\#/\#MLVSS{\cdot}day$
Domestic	0.73	0.52	0.075
Refinery	0.49–0.62	0.40–0.77	0.10–0.16
Chemical and Petrochemical	0.31–0.72	0.31–0.76	0.05–0.18
Brewery	0.56	0.48	0.10
Pharmaceutical	0.72–0.77	0.46	--
Kraft Pulping and Bleaching	0.5	0.65–0.8	0.08

*All data include the effect of influent suspended solids

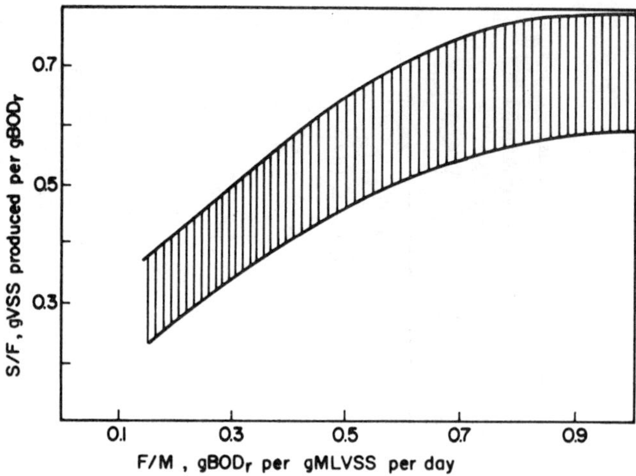

Figure 2-7. Sludge production relationship.

The oxygen requirement associated with the substrate uptake and the sludge synthesis can be presented in the form of the simplified equation

$$\Delta O_2 = a^1 S_r + b^1 X_v \tag{2-21}$$

where: ΔO_2 = oxygen requirement in lb/day (or g/day), a^1 = coefficient of oxygen requirement, and b^1 = 1.42b. The observed "a^1" values are also presented in Table 2-2.

In practice, oxygen requirements differ for different substrates. They can be expressed by the "respiratory quotient" (RQ) which is the ratio of moles CO_2 produced per mole O_2 used. The respiratory quotient for carbohydrates is equal to 1.0, for proteins - 0.8, and 0.7 for fats. Retention of particulate matter and storage of substrates in bacterial cells also affect the values of the respiratory quotient.

2.9 NITROGEN TRANSFORMATIONS

Nitrogen compounds may occur in wastewater in the form of organic nitrogen, ammonia, nitrite and nitrate. In domestic sewage, normally only organic nitrogen and ammonia are present. During carbonaceous oxidation of wastewater, many forms of organic nitrogen are converted to ammonia (ammonification). In turn, further biological oxidation, mostly facilitated by proper

autotrophic micro-organisms (see p. 2.1), converts ammonia to nitrite
and nitrate (nitrification). Heterotrophic reduction of nitrite and nitrate
to molecular nitrogen (dissimilatory denitrification) leads to a substantial
elimination of nitrogen from wastewater. This mechanism of nitrogen elimination from wastewater helps to prevent eutrophication of receiving water
bodies, and to increase quality aspects of the respective water resources.
In another mechanism (assimilatory denitrification) nitrate is reduced to
ammonia level and is utilized in cell synthesis.

NITRIFICATION

The nitrification process is characterized by very substantial oxygen requirements and by decreases in wastewater alkalinity. The stoichiometric
oxygen requirement for nitrification is 4.57 gO_2/g of ammonia nitrogen
(see p. 2.1). This calculation excludes the role of micro-organisms synthesis in this process. Taking it into consideration will decrease the above
requirement by about 8 percent. The decrease in alkalinity for nitrification
is stoichiometrically equal to 7.14 g $CaCO_3$ per g of ammonia nitrogen oxidized. This value is only marginally affected by the biosynthesis accompany-
ing the process. The measured alkalinity losses fluctuate between 5.7 and
7.4 g $CaCO_3$ per g of ammonia nitrogen oxidized, and the difference between
these values and the stoichiometric value is attributed to the mineralization
of organic nitrogen (Scearce et al., 1980), and carbon dioxide stripping.
Such decreases in wastewater alkalinity may cause a drop in its pH values
when the alkalinity of the wastewater is low or its ammonia contents relatively high.

Although nitrification is possible at a relatively broad range of pH
values, the optimal results are reported mostly at pH of 7.5 to 8.5. Moderate decreases in pH are inhibitory but not toxic (Poduska and Andrews,
1974). It is also possible to acclimate nitrifying microflora to some pH
values which are not drastically different from the optimal values for the
specific application of the nitrification process. For low-ammonia wastewater, it is suggested that the nitrification tanks be operated at pH of 7.6
to 7.8, although the optimum pH for nitrification is found at higher values.
The former range of pH, however, allows the carbon dioxide produced to
escape to the atmosphere.

Knowledge of the influence of dissolved oxygen levels on the nitrification
process is as yet incomplete. Effective nitrification was found in systems
with dissolved oxygen contents of only 0.5 mg/L. However, in some experiments, it was observed that increased dissolved oxygen concentrations (up
to 8 mg/L) enhanced the process kinetics. All available information on the
subject is from studies of combined carbon oxidation-nitrification systems
while, in the separate stage nitrification systems, this relationship may be
different (Parker et al., 1975).

The mixed liquor temperature affects the rate of nitrification. This
dependence was studied in the range from 5° C to 30° C and a reasonable

agreement with the van't Hoff-Arhenius law was found. Also some concentrations of ammonia, nitrite and nitrate influence the process. Anthonisen et al. (1974) indicated that free ammonia is more inhibitory to <u>Nitrobacter</u> than to <u>Nitrosomonas</u>. It was found also that inhibitory concentrations of free ammonia and free nitrous acid are a function of temperature, the number of nitrifying organisms, and their prior exposure to the non-inhibitory concentration of these inhibitory factors. Heavy metals are strong inhibitors to the nitrification process when they are present in ionic form. However, at optimal pH values for nitrification (pH > 7.5) these metals are relatively insoluble and do not affect the process at levels of 10 to 20 mg/L (Parker et al., 1975).

DENITRIFICATION

A broad range of heterotrophic bacteria can accomplish denitrification under anoxic conditions (absence of dissolved oxygen). Various substrates (municipal sewage, brewery wastewater, molasses, etc.) may be used as electron donors for this reaction (McCarthy et al., 1969, and Monteith et al., 1980). In practical applications, however, methanol (CH_3OH) is often used for this purpose because of a low yield of biomass and an attractive cost in the past. The respective reaction is as follows:

$$NO_3^- + 0.83\ CH_3OH \rightarrow 0.5N_2 + 1.17\ H_2O + 0.83\ CO_2 + OH^- \quad (2-22)$$

The methanol requirement can be expressed by an equation developed by McCarthy et al. (1969):

$$C_m = 2.47\ NO_3^-\text{-N} + 1.53\ NO_2^-\text{-N} + 0.87\ DO \quad (2-23)$$

where C_m is the methanol dosage, mg/L; NO_3-N is the nitrate nitrogen removed, mg/L; NO_2^--N is the nitrite nitrogen removed, mg/L; and DO is the dissolved oxygen originally present, mg/L. The theoretical methanol dosage, ignoring biosynthesis, is 1.9 g of methanol per g of nitrate nitrogen. Including synthesis this requirement is increased to 2.47. The experimental data are in the range of 3 g/g.

In the denitrification process, alkalinity of wastewater is increased at a stoichiometric rate of 3.57 g as $CaCO_3$ per g of nitrate or nitrite nitrogen reduced to nitrogen gas. This offsets by one-half the alkalinity loss caused by nitrification. The experimental measurements of alkalinity increased in the denitrification process show values slightly lower than 3 g $CaCO_3$ per gram of nitrate or nitrite nitrogen reduced. The difference is attributed to the oversimplification of stoichiometric equations.

The denitrification rates are depressed below pH 6.0 and above pH 8.7 and the highest rates are within the range pH 7.0 to 7.5 (Parker et al., 1975). Denitrification is temperature dependent, but is technically feasible even at temperatures as low as 5° C (Murphy and Dawson, 1972; and Sutton et al., 1975).

An interesting possibility exists of using elemental sulfur in an autotrophic process as a substitute for methanol in the denitrification process. However, present knowledge is inadequate to evaluate this process (Batchelor and Lawrence, 1978).

2.10 PHOSPHORUS TRANSFORMATIONS

High levels of phosphorus contents in sewage treatment plant effluents have been held largely responsible for the increase in the eutrophication (over-fertilization) of surface water bodies. The main sources of phosphorus in sewage are: human waste (30-50%), and phosphate builders in detergents (50-70%). In industrial wastewater field, phosphorus content presents a problem in food processing and canning effluents, effluents containing poly-phosphates used to prevent corrosion and scale formations, effluents from some fertilizer plants, etc. Human excretions contain from 0.5 to 2.3 lb P/cap/yr (from 226 to 1004 g P/cap/yr), and phosphorus load in detergent builders is estimated at 2.3 lb P/cap/yr (1004 g P/cap/yr). The concentration of phosphorus in raw sewage ranges up to about 10 mg/L P, and the requirements for its removal - depending on the type of receiving water body - calls for maximal concentrations lower than 0.1 to 2.0 mg/L P.

PHOSPHORUS UPTAKE AND RELEASE

In activated sludge treatment of wastewater, it was observed (Wells, 1969) that at similar BOD removal efficiencies in the range from 90 to 95 percent, the phosphate removal rate could vary from 1 to 65 mg/L/hr. The phosphate acclimatized sludges always contained more phosphorus (6 to 8 percent P on a dry weight basis) than the non-acclimatized ones (3 to 4 percent P). About half of the total phosphates were tightly bound (released only by vigorous acid treatment) by these sludges and the rest could migrate freely back and forth from the liquid to the solid phase.

Randall et al. (1970) found that ortho-phosphate release occurred under extended aeration conditions of the activated sludge, and that the rate of this release was directly related to the magnitude of the phosphate uptake prior to release. They also observed that anoxic conditions caused a rapid release of soluble phosphate occurring mostly within 90 minutes after anoxic conditions had been reached. Also Brar and Tollefson (1975) demonstrated that phosphate removals in excess of growth requirements can occur in activated sludge treatment.

PHOSPHORUS LUXURY UPTAKE

The mechanisms of the excessive phosphorus removal by activated sludge
were explained by Levin and Shapiro (1965) as being due to the "luxury uptake" by the micro-organisms. Menar and Jenkins (1969) understood the
phenomenon only as a chemical precipitation associated with the rise in pH
value of the mixed liquor due to the stripping of CO_2. However, Sherrard
and Schroeder (1972) saw in it only a normal assimilation of phosphates.

Several studies were carried out to clarify the above differences of
opinion. Morgan and Fruh (1974) observed that the maximum activated
sludge phosphorus content for normal growth was approximately one percent.
In addition, a storage zone existed to incorporate phosphorus to a content of
about 1.6 percent, with very effective utilization of influent phosphorus. At
a higher influent phosphorus concentration, the sludge could incorporate over
3 percent of phosphorus but not all phosphorus would be utilized. The findings that the phosphorus "luxury uptake" was a general rule in the activated
sludge process were confirmed in the work of Stall and Sherrard (1976).
Moreover, they found that phorphorus removal efficiencies were directly related to sludge production, and that decreasing aeration time resulted in increasing phosphorus removals. Buchan (1981) used electron microscopy,
combined with the energy dispersive analysis of X-rays, to examine the
nature of phosphorus accumulated in some activated sludges. The phosphorus was located in large electron-dense bodies, containing in excess of 30% P,
and present within large bacterial cells which were characteristically grouped
in clusters. The calcium to phosphorus ratio of these electron-dense bodies
did not resemble any form of calcium phosphate precipitate. These results
support the concept of a biological mechanism of enhanced phosphorus uptake
in the activated sludge process.

Barnard (1975) achieved removal of phosphorus from sewage with initial
concentrations between 9 and 12 mg/L P, to concentrations below 1 mg/L P,
by treatment in a special modification of the activated sludge process designed for removal of nutrients without any addition of chemicals. It was
observed that removal of phosphorus was positively influenced by an effective
removal of nitrogen, and was adversely affected by increasing sludge age
(sludge retention time). In this method, phosphorus leaves the system with
the excess of activated sludge and has to be released and further processed
by chemical precipitation.

The technical applications of phosphorus removal in various modifications of the activated sludge process are described in Chapter IV.

2.11 ELIMINATION OF PATHOGENIC BACTERIA AND VIRUSES

It is well accepted that the activated sludge process is very effective in removing bacteria and viruses from wastewater, including their pathogenic
forms. However, the particular modifications of the process (see Chapter
IV) may differ substantially in their respective effectiveness.

TABLE 2-3
Removal of Organisms by Conventional
Activated Sludge System
(after: Sproul, 1978)

Organisms	Percent Removal
Bacteria:	
E. typhosa	86- 99
Cholera	96-100
Tubercle bacilli	90+
Coliform	97
Fecal streptococci	96
Viruses:	
Coxsackie A9	96-99
Polio 1, 2, 3	76-90
Mixed	53-71
Parasites:	
Tapeworm eggs	0
E. histolytica cysts	incomplete

A measure of concentration of bacteria in wastewater can be presented as a standardized bacteria enumeration or as a statistical estimate like MPN (most probable number) index. These values for the enteric bacteria (Coliform group, fecal stroptococci, etc.) serve as indicators of fecal pollution. The latter information is assumed to be an indirect measure of the probability of the presence of pathogenic organisms in wastewater.

In activated sludge treatment, the bacterial composition of wastewater undergoes a drastic change (Table 2-3). In several studies, it has been observed that the contents in the sewage of E. coli, fecal streptococci, indoleforming, ammonifying, etc. bacteria were considerably reduced. It is believed that pathogenic bacteria and enertic viruses are effectively eliminated in such treatment except with some short-time aeration modifications of the process. However, the effect of the treatment on cysts of parasitic protozoa and worm ova is usually not great.

Viruses are electrically charged colloidal particles ranging in size from less than 30 to 300 mμ, which may adsorb to surfaces outside the host cells. The sorptive properties have a strong influence on the behaviour of viruses in the activated sludge process. Over 100 types of enteric viruses have been shown to be actually or potentially waterborne, and most of the enteric viruses in surface waters are probably the result of inadequately treated sewage. Infectious hepatitis is the most well-known viral disease, and, perhaps, viral gastroenteritis is the most common one.

The virus binding and inactivating capacity of activated sludge treatment has been studied by numerous authors. It was generally found that such treatment decreased the virus content in sewage by 1.5 log units; however, the process was sometimes much less effective (Lund et al., 1969). Malina et al. (1975) performed some studies on the mechanism of polio virus inactivation by activated sludge. It was found that, after an initial adsorption of virus in the range of 1.6 x 10^9 to 1.9 x 10^9 PFU per g of sludge (PFU - infectivity in plaque-forming units), no appreciable further inactivation took place during the aeration period. The virus eluted from the sludge showed much smaller specific infectivity than the original material. Data collected from a seeded pilot-grant of a three-sludges modification of the activated sludge process (see Chapter IV) showed a combined viral removal capacity of 99.9 percent (Safferman and Morris, 1976). Glass and O'Brien (1980) emphasized that enterovirus removal by activated sludge cannot be equated with virus inactivation, and the concentration of viruses in the excess activated sludge makes the sludge potentially hazardous material.

REFERENCES

Adams, C. E. et al.: "A Kinetic Model for Design of Completely Mixed Activated Sludge Treating Variable-Strength Industrial Wastewater", Water Research (Brit.), 9, 37, (1975).

Angelbeck, D. I., and Kirsch, E. J.: "Influence of pH and Metal Cations on Aggregative Growth of Non-Slime Forming Strains of Zooglea ramigera", Appl. Microbiol., 17, 435 (1969).

Anthonisen, A. C.: "The Effects of Free Ammonia and Free Nitrous Acid on the Nitrification Process", Ph.D. Thesis, Cornell University, Ithaca, N.Y. (1974).

Baillod, C. R., and Boyle, W. C.: "Mass Transfer Limitations in Substrate Removal", Proc. Amer. Soc. Civ. Engrs., Jour. San. Eng. Div., 96, SA2, 525, (1970).

Barnard, J. L.: "Biological Nutrient Removal Without the Addition of Chemicals", Water Research (Brit.), 9, 485 (1975).

Batchelor, B., and Lawrence, A. W.: "A Kinetic Model for Autotrophic Denitrification Using Elemental Sulfur", Water Research (Brit.), 12, 1075, (1978).

Benedict, A. H. et al.: "Sludge Production, Waste Composition, and Biochemical Oxygen Demand Loading Effects for Activated Sludge Systems", Jour. Water Poll. Control Fed., 51, 2898, (1979).

Bishop, D. F., et al.: "Single Stage Nitrification - Denitrification", paper presented at the 47th Annual Conference of Water Poll. Control Fed., Denver, Colorado, 1974.

Boyle, W. C. et al.: "Flocculation Phenomena in Biological Systems", from 'Advances in Water Quality Improvement, Water Resources Symposium No. 1'. Edited by E. F. Gloyna and W. W. Eckenfelder, Jr., University of Texas Press, Austin, Texas, 1968.

Brar, G. S., and Tollefson, E. L.: "The Luxury Uptake Phenomenon for Removal of Phosphates from Municipal Wastewater", Water Research (Brit.), 9, 71, (1975).

Breed, R. S. et al.: "Bergey's Manual of Determinative Bacteriology", 7th ed., The Williams and Wilkins Company, Baltimore, Md., 1957.

Brezonik, P. L., and Patterson, J. W.: "Activated Sludge ATP. Effects of Environmental Stress", Proc. Am. Soc. Civ. Engrs., Jour. San. Eng. Div., 97, 813, (1971).

Buchan, L.: "The Location and Nature of Accumulated Phosphorus in Seven Sludges from Activated Sludge Plants which Exhibited Enhanced Phosphorus Removal", Water SA, 7, 1, (1981).

Buchanan, R. E.: "Agglutination", Jour. Bacteriol., 4, 71, (1919).

Bukantz, S. C. et al.: "The Elaboration of Soluble Capsular Polysaccharide by Pneumococcus III in Relation to Growth Phases in Vitro", Jour. Bacteriol., 42, 29, (1941).

Burkhead, C. E., and McKinney, R. E.: "Energy Concepts of Aerobic Microbial Metabolism", Proc. Amer. Soc. Civ. Engrs., Jour. San. Eng. Div., 95, SA2, 267, (1969).

Burkhead, C. E., and Waddell, S. L.: "Composition Studies of Activated Sludges", Proc. 24th Ind. Waste Conf., Purdue Univ., Ext. Ser. 135, 576, (1970).

Busch, P. L., and Stumm, W.: "Chemical Interactions in the Aggregation of Bacteria", Environmental Science and Technol., 2, 49, (1968).

Butterfield, C. T.: "Studies of Sewage Purification; a Zooglea-forming Bacterium Isolated from Activated Sludge", U.S.P.H.S. Publ. Health Reps., 50, Part I, 671, (1935).

Catchpole, J. R., and Cooper, R. L.: "The Biological Treatment of Carbonization Effluents III. New Advances in the Biochemical Oxidation of Liquid Wastes", Water Research (Brit.) 6, 1459 (1972).

Carr, D. F., and Ganczarczyk, J.: "Activated Sludge Exocellular Material - Extraction Methods and Problems", Proc. 9th Canadian Symposium on Water Pollution Research, pp. 250-261 (1974).

Cheng, M. H. et al.: "Heavy Metals Uptake by Activated Sludge," Jour. Water Poll. Control Fed., 40, 362, (1975).

Cooke, W. G., and Pipes, W. O.: "The Occurrence of Fungi in Activated Sludge," Proceedings 23rd Industrial Waste Conf., Purdue University Extension Serv., 132, 177, (1969).

Crabtree, K. et al.: "Isolation, Identification and Metabolic Role of the Sudanophilic Granules of Zoogloea ramigera," Appl. Microbiol., 13, 218, (1965).

Crabtree, K. et al.: "A Mechanism of Floc Formation by Zoogloea ramigera," Jour. Water Poll. Control Fed., 38, 1968, (1966).

Crabtree, K., and McCoy, E.: "Zoogloea ramigera Itzigsohn, Identification and Description," Intern. Jour. Systematic Bacteriol., 17, 1, (1967).

Curds, C. R., and Cockburn, A.: "Protozoa in Biological Sewage-Treatment Processes", Water Research (Brit.), 4, 225 & 237 (1970).

Davies, D. A. L.: "The Specific Polysaccharides of Some Gram-negative Bacteria," Biochem. Jour., 59, 696, (1955).

Dean, R. B.: "Colloids Complicate Treatment Processes", Environ. Science and Technol., 3, 820, (1969).

DeWalle, F. B., and Chian, E. S. K.: "Tertiary Treatment of Secondary Effluent for Potential Reuse," paper presented at the 47th Annual Conf. of Water Poll. Control Fed., Denver, Col., October 1974.

DeWalle, F. B. et al.: "Organic Matter Removal by Powdered Activated Carbon Added to Activated Sludge Units," paper presented at 48th Annual Conf. of Water Poll. Control Fed., Miami Beach, Fla., October 1975.

Dias, F. F., and Bhat, J. V.: "Accumulation of PHB and Iodophilic Material by the Dominant Activated Sludge Bacteria," Current Science, 32, 501 (1963).

Dias, F. F., and Bhat, J. V.: "Microbial Ecology of Activated Sludge. I. Dormant Bacteria," Appl. Microbiol., 12, 412 (1964).

Dixon, J. K. et al.: "Effect of the Structure of Cationic Polymers on the Flocculation and the Electrophoretic Mobility of Crystalline Silica," Jour. Coll. Interface Science, 23, 465 (1967).

Dixon, J. K., and Zielyk, M. W.: "Control of the Bacterial Content of Water with Synthetic Polymeric Flocculents," Environ. Science and Technol., 3, 551 (1969).

Dudman, W. F., and Wilkinson, J. F.: "The Composition of the Extra-cellular Polysaccharides of Aerobacter-Klebsiella Strains," Biochem. Jour., 62, 289 (1956).

Dugan, P. R., and Lundgren, D. G.: "Isolation of the Floc-forming Organism Zoogloea ramigera and Its Culture in Complex and Synthetic Media," Appl. Microbiol., 8, 357 (1960).

Duguid, J. P., and Wilkinson, J. F.: "The Influence of Cultural Conditions on Polysaccharide Production by Aerobacter aerogenes," Jour. Microb. 9, 174 (1953).

Eckenfelder, W. W. Jr.: "Thermodynamics of Biological Synthesis and Growth," discussion in "Advances in Water Poll. Res.," Vol. 2, J. K. Baars (Ed.), Pergamon Press, New York, N. Y., p. 165 (1965).

Eckenfelder, W. W., Jr.: "Water Quality Engineering for Practicing Engineers," Cahners, Boston, 1970.

Eikelboom, D. H.: "Filamentous Organisms Observed in Activated Sludge," Water Research (Brit.), 9, 365 (1975).

Farquhar, G. J., and Boyle, W. C.: "Identification of Filamentous Micro-organisms in Activated Sludge," Jour. Water Poll. Control Fed., 43, 4, 604 (1971).

Farquhar, G. J., and Boyle, W. C.: "Control of Thiothrix in Activated Sludge," Jour. Water Poll. Control Fed., 44, 1, 14 (1972).

Ford, D. L., and Eckenfelder, W. W.: "Effect of Process Variables on Sludge Floc Formation and Settling Characteristics," Jour. Water Poll. Control Fed., 39, 1850 (1967).

Friedman, B. A., and Dugan, P. R.: "Identification of Zoogloea Species and the Relationship to Zoogloeal Matrix and Floc Formation," Jour. Bacteriol., 95, 1903 (1968).

Friedman, B. A. et al.: "Fine Structure and Composition of the Zoogloeal Matrix Surrounding Zoogloea ramigera," Jour. Bacteriol., 96, 2144 (1968).

Friedman, B. A. et al.: "Structure of Exocellular Polymers and Their Relationship to Bacterial Flocculation," Jour. Bacteriol., 98, 1328 (1969).

Ganczarczyk, J.: "Nitrogen Transformations in Activated Sludge Treatment," Proc. Amer. Soc. Civ. Engrs., Jour. Sanit. Eng. Div., 97, SA3, 247 (1971).

Ganczarczyk, J., and Obiaga, T.: "Mechanism of Lignin Removal in Activated Sludge Treatment of Pulp Mill Effluents," Water Research (Brit.), 8, 857 (1974).

Garret, M. T., Jr., and Sawyer, C. N.: "Kinetics of Removal of Soluble BOD by Activated Sludge," paper presented at the 7th Purdue Ind. Waste Conf., Lafayette, Ind., 1952.

Gaudy, A. F., Jr.: "Studies on Induction and Repression in Activated Sludge Systems," Appl. Microbiol., 10, 264 (1962).

Gaudy, A. F., Jr., and Engelbrecht, R. S.: "Basic Biochemical Considerations during Metabolism in Growing vs. Respiring Systems," Advances in Biological Waste Treatment, Pergamon Press, London, 1963.

Gaudy, A. F., Jr., et al.: "Sequential Substrate Removal in Heterogeneous Populations," Jour. Water Poll. Control Fed., 35, 903 (1963).

Gaudy, A. F., Jr. et al.: "Activated Sludge Process Modification for Nitrogen Deficient Wastes," presented at the 4th International Conf. on Water Poll. Research, Section II, Pergamon Press, London, 1968.

Gaudy, A. F., Jr. et al.: "Continuous Oxidative Assimilative of Acetic Acid and Endogenous Protein Synthesis Applicable to Treatment of Nitrogen-Deficient Waste Water," Jour. Appl. Microbiol., 16, 1358 (1968).

Ghosh, S. et al.: "Phasic Utilization of Substrates by Aerobic Cultures," Jour. Water Poll. Control Fed., 44, 3, 376 (1972).

Glass, J. S., and O'Brien, R. T.: "Enterovirus and Coliphage Inactivation During Activated Sludge Treatment," Water Research (Brit.), 14, 877 (1980).

Goel, K. C., and Gaudy, A. F., Jr.: "Regeneration of Oxidative Assimilation Capacity by Intracellular Conversion of Storage Products to Protein," Jour. Appl. Microbiol., 16, 1352 (1968).

Haltrich, W.: "Influence of Nitrogen Supply on Biological Treatment," paper presented at the 24th Purdue Industrial Waste Conference, Lafayette, Ind., 1969.

Harris, R. H., and Mitchell, R.: "Inhibition of the Flocculation of Bacteria by Biopolymers," paper presented at the Annual Meeting of the Water Pollution Control Fed., Atlanta, Georgia, 1972.

Hattingh, W. H. J.: "Activated Sludge Studies the Nitrogen and Phosphorus Requirements of the Micro-organisms," Water and Waste Treatment (Brit.), 9, No. 8, 380 (1963).

Healy, T. W., and LaMer, V. K.: "Adsorption - Flocculation Reactions of a Polymer with an Aqueous Colloidal Dispersion," Jour. Physical Chem., 66, 1835 (1962).

Heller, W., and Pugh, T. L.: "Steric Stabilization of Colloidal Solutions by Adsorption of Flexible Macromolecules," Jour. Polymer Sci., 47, 149, 203 (1960).

Helmers, E. N. et al.: "Nutritional Requirements in the Biological Stabilization of Industrial Wastes, III, Treatment with Supplementary Nutrients," Sewage and Ind. Wastes, 24, 496 (1952).

Heukelekian, H., and Littman, M. L.: "Carbon and Nitrogen Transformations in the Purification of Sewage by the Activated Sludge Process. II. Morphological and Biochemical Studies of Zoogleal Organisms," Sewage Wks. Jour., 11, 752 (1939).

Heukelekian, H., et al.: "Factors Affecting the Quantity of Sludge Production in the Activated Sludge Process," Sewage and Ind. Wastes, 23, 945 (1951).

Hodge, H. M., and Metcalfe, S. N., Jr.: "Flocculation of Bacteria by Hydrophilic Colloids," Jour. Bacteriol., 75, 258 (1958).

Hoover, S. R., and Porges, N.: "Assimilation of Dairy Wastes by Activated Sludge. II. The Equation of Synthesis and Rate of Oxygen Utilization," Sewage and Ind. Wastes, 24, 306 (1952).

Jewell, W. J., and Cummings, R. J.: "Denitrification of Concentrated Wastewater," Jour. Water Poll. Control Fed., 47, 2281 (1975).

Jones, P. H.: "The Effect of Nitrogen and Phosphorus Compounds on One of the Microorganisms Responsible for Sludge Bulking," Proc. 20th Annual Industrial Waste Conference, Purdue University, Lafayette, Ind., 1965.

Jones, P. H., and Prasad, D.: "The Use of Terazolium Salts as a Measure of Sludge Activity", Jour. Water Poll. Control Fed., 41, R441 (1969).

Komolrit, K. et al.: "Regulation of Exogenous Nitrogen Supply and its Possible Applications to the Activated Sludge Process," Jour. Water Poll. Control Fed., 39, 251 (1967).

LaMer, V. K. et al.: "Flocculation, Subsidence and Filtration of Phosphate Slimes. Part IV. Flocculation by Gums and Polyelectrolytes and their Influence on Filtration Rate," Jour. Colloid Sci., 12, 230 (1957).

LaMer, V. K., and Healy, T. W.: "Adsorption - Flocculation Reactions of Macromolecules at the Solid-Liquid Interface," Reviews of Pure and Applied Chemistry, 13, 112 (1963).

Lawrence, A. W., and McCarty, P. L.: "Unified Basis for Biological Treatment Design and Operation," Proc. Am. Soc. Civ. Engrs., Jour. San. Eng. Div., 96, SA3, 757 (1970).

Levin, G. V., and Shapiro, J.: "Metabolic Uptake of Phosphorus by Wastewater Organisms," Jour. Water Poll. Control Fed., 37, 800 (1965).

Lighthart, B., and Oglesby, R. T.: "Bacteriology of an Activated Sludge Wastewater Treatment Plant. A Guide to Methodology," Jour. Water Poll. Control Fed., 41, 8/2, R267 (1969).

Lund, E. et al.: "Occurrence of Enteric Viruses in Wastewater after Activated Sludge Treatment," Jour. Water Poll. Control Fed., 41, 169 (1969).

MacRae, R. M., and Wilkinson, J. R.: "Poly-beta-hydroxybutyrate Metabolism in Washed Suspensions of Bacillus cereus and Bacillus megaterium," Jour. Gen. Microbiol., 19, 210 (1958).

Malina, J. R., Jr. et al.: "The Mechanism of Polio Virus Inactivation by Activated Sludge," Jour. Water Poll. Control Fed., 47, 2178 (1975).

Matson, J. V. et al.: "Oxygen Supply Limitations in Full Scale Biological
Treatment Systems," paper presented at the 27th Ind. Waste Conf., Purdue
Univ., Lafayette, Ind., 1972.

Matson, J. V., and Characklis, W. G.: "Diffusion into Microbial Aggre-
gates," Water Research (Brit.), 10, 877 (1976).

McCarty, P. L.: "Thermodynamics of Biological Synthesis and Growth,"
in "Advances in Water Poll. Res.," Vol. 2, J. K. Baars (Ed.), Pergamon
Press, New York, N. Y., p. 169 (1965).

McCarty, P. L., Beck, L., and St. Amant, P.: "Biological Denitrification
of Wastewaters by Addition of Organic Materials," Proc. 24th Ind. Waste
Conference, Purdue University, Lafayette, Ind. May 1969.

McGregor, W. C., and Finn, R. K.: "A Study of Microbial Floc Formation
by Using Continuous Recording of Optical Density," Chemical Eng. Symp.
Series, 67, 108, 152 (1972).

McKinney, R. E., and Horwood, M. P.: " A Fundamental Approach to the
Activated Sludge Process. I. Floc-Producing Bacteria," Sewage Ind.
Wastes, 24, 117 (1952).

McKinney, R. E.: "A Fundamental Approach to the Activated Sludge Process. II. A Proposed Theory of Floc Formation," Sewage Ind. Wastes,
24, 280 (1952).

McKinney, R. E., and Weichlein, R. G.: "Isolation of Floc-Producing
Bacteria from Activated Sludge," Appl. Microbiol., 1, 259 (1953).

McKinney, R. E., and Gram, A.: "Protozoa and Activated Sludge,"
Sewage Ind. Wastes, 28, 1219 (1956).

Menar, A. B., and Jenkins, D.: "The Fate of Phosphorus in Activated
Sludge Treatment," Proc. 24th Ind. Waste Conf., Purdue Univ., Lafayette,
Ind., May, 1969.

Monod, J.: "The Growth of Bacterial Cultures," Annual Review of Microbiology, 3, 371 (1949).

Monteith, H. D., et al.: "Industrial Waste Carbon Sources for Biological
Denitrification," paper presented at the Xth Conference of Int. Assoc. on
Water Poll. Res., Toronto, 1980.

Morgan, W. E., and Fruh, E. G.: "Phosphate Incorporation in Activated
Sludge," Jour. Water Poll. Control Fed., 46, 2486 (1974).

Mueller, J. A. et al.: "Nominal Diameter of Floc Related to Oxygen
Transfer," Proc. Am. Soc. Civ. Engrs., Jour. San. Eng. Div., 92, SA2,
4756 (1966).

Murphy, R. L., and Dawson, R. N.: "The Temperature Dependency of
Biological Denitrification," Water Research (Brit.), 6, 71 (1972).

Neufeld, R. D., and Hermann, E. R.: "Heavy Metals Removal by Accli- mated Activated Sludge," Jour. Water Poll. Control Fed., 47, 310 (1975).

Painter, H. A.: "A Review of Literature on Inorganic Nitrogen Metabolism in Microorganisms," Water Research (Brit.), 4, 393 (1970).

Painter, H. A., Denton, R. S., and Quarmby, C.: "Removal of Sugars by Activated Sludge," Water Research (Brit.), 2, 427 (1968).

Parker, D. S. et al.: "Process Design Manual for Nitrogen Control," Technology Transfer, Environmental Protection Agency, October 1975.

Parkin, G. F., and McCarty, P. L.: "Characteristics and Removal of Soluble Organic Nitrogen in Treated Effluents," paper presented at the 7th Internl. Conf. on Water Poll. Research, Paris, 1974.

Pasveer, A.: "Research on Activated Sludge. III. Distribution of Oxygen in Activated Sludge Floc," Sewage Ind. Wastes, 26, 28 (1954).

Pasveer, A.: "A Case of Filamentous Activated Sludge," Jour. Water Poll. Control Fed., 41, 1340 (1969).

Pavoni, J. L. et al.: "Bacterial Exocellular Polymers and Biological Flocculation," Jour. Water Poll. Control Fed., 44, 414 (1972).

Peter, G., and Wuhrmann, K.: "Contribution to the Problem of Biofloccu- lation in the Activated Sludge Process," from "Advances in Water Pollu- tion Research," Vol. 1, II-1, Proc. 5th International Water Pollution Research Conference. Edited by S. H. Jenkins, Pergamon Press, Oxford, England, 1971.

Pike, E. B., and Curds, C. R.: "The Microbial Ecology of the Activated Sludge Process," from Microbial Aspects of Pollution. The Society for Applied Bacteriology, Symposium Series No. 1. Academic Press 123, 1971.

Poduska, R. A., and Andrews, J. F.: "A Study of the Dynamics of Nitri- fication in the Activated Sludge Process," paper presented at the 47th Annual Conference of Water Poll. Control Federation, Denver, Colorado, 1974.

Prakasam, T. B. S., and Dondero, N. C.: "Aerobic Heterotrophic Bac- terial Populations of Sewage and Activated Sludge," Appl. Microbiol., 15, 461 and 1122 (1967).

Ramarao, C. V. et al.: "Treatment of Nitrogen-Deficient Organic Wastes by a Modification of the Activated Sludge Process," Proc., 18th Industrial Waste Conf., Purdue University, Ext. Ser. Vol. 115, 427 (1964).

Randall, C. W. et al.: "Phosphate Release in Activated Sludge Process," Proc. Amer. Soc. Civ. Engrs., Jour. San. Eng. Div., 96, SA2, 395 (1970).

Ries, H. E., Jr., and Meyers, B. L.: "Flocculation Mechanism, Charge Neutralization and Bridging," Science, 160, 3835, 1449 (1968).

Rouf, M. A., and Stokes, J. L.: "Isolation and Identification of the Sudanophilic Granules of Sphaerotilus natans," Jour. Bacteriol., 83, 343 (1962).

Safferman, R. S., and Morris, M. E.: "Assessment of Virus Removal by a Multi-Stage Activated Sludge Process," Water Research (Brit.), 10, 413 (1976).

Sandford, P. A., and Conrad, H. E.: "The Structure of the Aerobacter aerogenes A3 (S1) Polysaccharide. I. A Re-examination Using Improved Procedures for Methylation Analysis," Biochemistry, 5, 1508 (1966).

Sato, T., and Ose, Y.: "Floc-Forming Substances Extracted from Activated Sludge by Sodium Hydroxide Solutions," Water Research (Brit.), 14, 333 (1980).

Sawyer, C. N.: "Bacterial Nutrition and Synthesis," in "Biological Treatment of Sewage and Industrial Wastes," Vol. 1, Reinhold, New York, 1956.

Schnitzer, M.: "Fulvic Acid: A Possible Anti-Pollution Sponge," Chemistry in Canada, 25, 40 (1973).

Schroeder, E. D., and Busch, A. W.: "Mass and Energy Relationships in Aerobic Digestion," Proc. Amer. Soc. Civ. Engrs., Jour. San. Eng. Div., 92, SA1, 85 (1966).

Scearce, S. N., et al.: "Prediction of Alkalinity Changes in the Activated Sludge Process," Journ. Water Poll.,Control Fed., 52, 399 (1980).

Servizi, J. A., and Bogan, R. H.: "Free Energy as a Parameter in Biological Treatment," Proc. Amer. Soc. Civ. Engrs., Jour. Sanit. Eng. Div., 89, SA3 (1963).

Servizi, J. A., and Bogan, R. H.:"Thermodynamic Aspects of Biological Oxidation and Synthesis," Jour. Water Poll. Control Fed., 36, 5 (1964).

Sherrard, J. H., and Schroeder, E. D.: "Importance of Cell Growth Rate and Stoichiometry to the Removal of Phospnorus from the Activated Sludge Process," Water Research (Brit.), 6, 1059 (1972).

Singer, P. C. et al.: "Flocculation of Bulked Activated Sludge with Polyelectrolytes," Jour. Water Poll. Control Fed., 40, R1 (1968).

Speece, R. E. et al.: "Cell Replication and Biomass in the Activated Sludge Process," Water Research (Brit.), 7, 361 (1973).

Sproul, O. J.: "The Efficiency of Wastewater Unit Processes in Risk Evaluation," Proc. of the Conference on Risk Assessment and Health Effects of Land Application of Municipal Wastewater and Sludges, The University of Texas at San Antonio, 1978.

Stall, T. R., and Sherrard, J. H.: "Effects of Wastewater Composition and Cell Residence Time on Phosphorus Removal in Activated Sludge," Jour. Water Poll. Control Fed., 48, 307 (1976).

Stumm-Zollinger, E.: "Substrate Utilization in Heterogeneous Bacterial Communities," Jour. Water Poll. Control Fed., 40, R213 (1968).

Stumm, W., and Morgan, J.: "Aquatic Chemistry," Wiley Interscience, New York, 1970.

Sutton, P. M. et al.: "Low Temperature Biological Denitrification of Wastewater," Jour. Water Poll. Control Fed., 47, 122 (1975).

Swilley, E. L. et al.: "Significance of Transport Phenomena in Biological Oxidation Processes," Proc. 19th Ind. Waste Conf., Purdue Univ., Lafayette, Ind., 1964.

Symons, J. M., and McKinney, R. E.: "The Biochemistry of Nitrogen in the Synthesis of Activated Sludge," Sewage and Ind. Wastest, 30, 874 (1958).

Tenney, M. W., and Stumm, W.: "Chemical Flocculation of Microorganisms in Biological Waste Treatment," Jour. Water Poll. Control Fed., 37, 1370 (1965).

Tenney, M. W. et al.: "Algal Flocculation with Synthetic Organic Polyelectrolytes," Appl. Microbiol., 18, 965 (1969).

Tezuka, Y.: "Magnesium Ion as a Factor Governing Bacterial Flocculation," Appl. Microbiol. 15, 1256 (1967).

Tezuka, Y.: "Cation Dependent Flocculation in a Flavo-bacterium Species Predominant in Activated Sludge," Appl. Microbiol., 17, 222 (1969).

Tischler, L. F., and Eckenfelder, W. W., Jr.: "Linear Substrate Removal in the Activated Sludge Process," paper presented at the 4th Internl. Conf. on Water Poll. Research, Prague, Czechoslovakia, 1969.

Van Gils, H. W.: "Bacteriology of Activated Sludge," Report No. 32, Research Institute for Public Health Engineering, T.N.O., The Hague, Holland, (1964).

Wallen, L., and Davis, E. N.: "Biopolymers of Activated Sludge," Envir. Sci. and Technol., 6, 161 (1972).

Walters, C. F., Engelbrecht, R. S., and Speece, R. E.: "Microbial Substrate Storage in Activated Sludge," Proc. Amer. Soc. Civ. Engrs., Jour. Sanit. Eng. Div., 94, SA2 (1968).

Washington, D. R., and Symons, J. M.: "Volatile Sludge Accumulation in Activated Sludge Systems," Jour. Water Poll. Control Fed., 34, 769 (1962).

Wattie, E.: "Cultural Characteristics of Zoogloeal-forming Bacteria Isolated from Activated Sludge and Trickling Filters," Sewage Wks. Jour., 15, 476 (1943).

Wells, W. N.: "Differences in Phosphate Uptake Rates Exhibited by Activated Sludges," Jour. Water Poll. Control Fed., 41, 765 (1969).

Watson, R. F., and Eckenfelder, W. W., Jr.: "Application of Biological Treatment to Industrial Wastes. I. Kinetics and Equilibria of Oxidative Treatment," Sewage and Ind. Wastest, 27, 802 (1955).

Wild, H. E. et al.: "Factors Affecting Nitrification Kinetics," Jour. Water Poll. Control Fed., 43, 1845 (1971).

Wood, D. K., and Tchobanoglous, G.: "Trace Elements in Biological Waste Treatment," Jour. Water Poll. Control Fed., 47, 1933 (1975).

Wuhrmann, K.: "Factors Affecting Efficiency and Solids Production in the Activated Sludge Process," in "Biological Treatment of Sewage and Industrial Wastes," Reinhold, New York, 1957.

Wuhrmann, K.: "Effect of Oxygen Tension on Biochemical Reactions in Sewage Purification Plants," in "Advances in Biological Waste Treatment," edited by W. W. Eckenfelder, Jr., and J. McCabe, Pergamon Press, Oxford, 1963.

Factors Affecting Activated Sludge Process

Numerous factors influence the performance of the activated sludge process. The most important of these factors, which are also called the "parameters" of the process, are variability in wastewater flow, wastewater quality, aeration time, characteristic aeration tank loadings, sludge loadings (food to micro-organism ratio values), maintained levels of mixed liquor suspended solids, aeration intensities, maintained sludge solids retention times, physical characteristics of sludge, characteristics of mixing and turbulence in aeration tanks, mixed liquor temperature, and influent concentration.

3.1 WASTEWATER FLOW AND QUALITY

The dynamic behaviour of the activated sludge process is caused by the fluctuating quality and quantity of the input (sewage or industrial wastewater) as well as by the resulting difficulties in maintaining the required stability of the process parameters. The characteristic non-uniformity of the sewage or the specific wastewater quality and quantity may be partially controlled by the design and operation of wastewater collection systems, and by application of separate equalization units (in-line or side-line). The former approach is based on the utilization of the storage capacity of the trunk and interceptor lines, thus sufficiently regulating the flow to the treatment plant in many cases. However, for this purpose quite extensive instrumentation and computerized operation are required (Leiser, 1974). In the latter approach, separate equilization tanks are capable of making the flow and quality of wastewater more uniform (Wallace, 1968; Smith et al., 1973; DiToro, 1974; Novotny and Englade, 1974; DiToro, 1975; Foess et al., 1975; Novotny and Stein, 1976). It is also possible to achieve some wastewater flow equalization by use of aeration tank volume (Speece and LaGrega,

1976). In general, equalization of wastewater prior to activated sludge
treatment is more important for some industrial effluents (see Chapter VIII)
than for municipal wastewater.

3.2 WASTEWATER AERATION TIME

Periods of wastewater aeration with activated sludge, which are also called
wastewater hydraulic detention (or retention) times in aeration tanks, were
historically the first to be acknowledged as an important process parameter
(Ardern and Lockett, 1914; and Bartow and Mohlman, 1915). Wastewater
aeration times are usually expressed in hours, are based on wastewater
flow only, and ignore the flow of the sludge recycle. Prior to the 1940s,
recommended aeration times for treatment of domestic sewage were in the
range of 4 to 10 hours, but for many practical applications these values
were empirically selected on the basis of the characteristics and strength
of wastewater, and the desired degree of treatment. Accordingly, longer
hydraulic detention times are suggested to increase the "buffering capacity"
of the system against qualitative or quantitative transient or shock loads
(see p. 3.1).

3.3 VOLUME AND SLUDGE LOADINGS

An early practice in the use of the activated sludge process was to quote
loading of units in terms of pounds (or grams) of BOD per unit of aeration
tank volume (volume loading or volumetric loading or space load). As substantially different levels of micro-organisms can be maintained in aeration
tanks, a more rational approach is to cite a respective food to micro-
organism ratio (F/M) in pounds (or grams) of BOD per pound (or gram) of
mixed liquor suspended solids (MLSS) or mixed liquor volatile suspended
solids (MLVSS). F/M is also called sludge loading or substrate loading.
The latter formulation was not introduced until the 1940s.

The low F/M ratios (up to 0.7g BOD/g MLSS/day) are applicable for
activated sludge process modifications associated with a high degree of
BOD removal and some ammonia nitrification; the higher F/M values characterize the partial activated sludge treatment or high dissolved oxygen
modifications of the process like the commercial oxygen process and deep-
shaft system (see Chapter IV). Perhaps the highest F/M value observed is
5.7 g BOD/g MLSS/day, reported by Kuslikis (1978) for a deep-shaft treatment of brewery wastewater. Table 3-1 shows a general relationship between the F/M ratio and the removed fraction of BOD load (Haseltine, 1955).
However, several treatment plants have a much better performance level due
to better control of a few of the process-limiting factors.

TABLE 3-1

Relationship between Sludge Loadings and
Process Efficiency (after Haseltine, 1955)

Sludge Loadings g BOD/g MLSS d	Percent of BOD Reduction
0.1	90
0.2	90
0.3	90
0.4	90
0.5	90
1.0	75
1.5	67
2.0	62
2.5	58
3.0	57
3.5	56
4.0	55
4.5	54
5.0	53

The selected F/M value can be recalculated into the volumetric load
value (A) according to the formula:

$$A = F/M \times MLVSS \tag{3-1}$$

Example: for F/M = 0.3 g BOD/g MLVSS/day, and MLVSS level of
2,500g/cu.m., the volumetric load will be: A = 0.3 2,500 = 750g
BOD/cu.m/day.

Some activated sludge treatment experiments indicate that the treatment results are a function of both solids retention time and hydraulic retention time, and that neither the volume loading nor the sludge loading
alone provide all the required information. Therefore, it is common to

report both values, in order to characterize fully the specified activated sludge process modification.

The response of the activated sludge process to time variations in hydraulic and/or organic loadings has been the subject of several studies (Eckhoff and Jenkins, 1966; Knopp et al., 1968; Duggan and Cleasby, 1976; Krishnan and Gaudy, 1976; Seleh and Gaudy, 1978, Selma and Schroeder, 1978; etc.). The mechanism of the response is complex and depends on the applied activated sludge process characteristics and the magnitude of variation.

3.4 ACTIVATED SLUDGE VOLUME INDEX

The activated sludge volume index (SVI), called also the Mohlman index (Mohlman, 1934) is the unit volume of the activated sludge (in mL/g) after a half an hour settling of the mixed liquor in a 1-liter Imhoff cone. The variations in the value of this index are the biological responses to many environmental factors, i.e., the sludge loading, wastewater constituents, nutrients demand, etc. The large number of these factors makes it difficult to predict the respective trends and values of SVI.

The values of the sludge volume index for various treatment cases are, generally, in the range of 40 to 300 mL/g. For practical needs this factor should be kept below 150 mL/g. With higher SVI values a difficulty arises in the performance of the secondary clarifiers. A distinct increase in the sludge volume index is called sludge bulking. The development of filamentous organisms (Smit, 1934; Ruchhoft and Kachmar, 1941; Jones, 1965) and the increase of zooglea bound water (Heukelekian and Weisberg, 1956) are considered causes of the activated sludge bulking. Both these factors are influenced by the activated sludge loading and the chemical composition of the wastewater treated (Ganettelli and Heukelekian, 1964).

One of the frequent causes of activated sludge bulking is excessive growth of filamentous micro-organisms (see Chapter II) in a mixed culture. This growth effectively competes with other forms of aerobic bacterial growth at low dissolved oxygen levels (below 2 mg/L), as it develops a larger surface per unit of bacterial mass. Its presence at moderate levels may be beneficial for the removal of residual BOD and suspended solids, but at excessive concentrations, separation of treated wastewaters and activated sludge by sedimentation becomes difficult.

To avoid activated sludge bulking, it is practical to keep the activated sludge BOD loadings within a proper range for a given modification of the treatment process and given wastewater characteristics. Various relationships between SVI values and activated sludge BOD loadings (F/M ratio) are given in Figure 3-1. This relationship was studied for sewage by Logan and Budd (1956). Eckenfelder (1967) based his studies on these data and his own (Ford and Eckenfelder, 1967), published a curve very similar to that of Ganczarczyk (1969) for Kraft pulp mill effluent. Both curves are practically

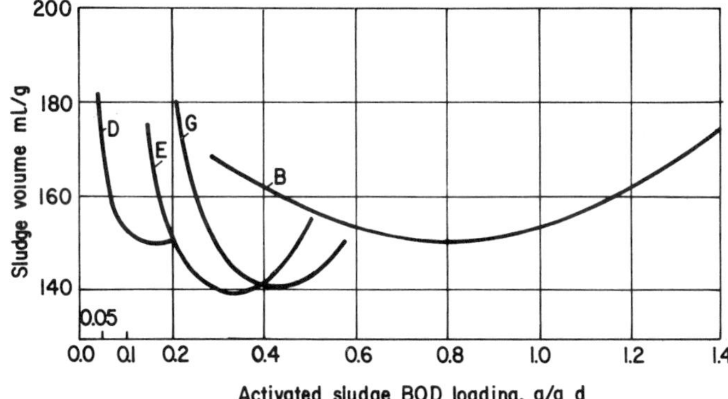

Figure 3-1. Comparison of various relationships between sludge volume
index values and activated sludge BOD loadings (D = after Dart and Spurr,
E = 1 after Eckenfelder, G = after Ganczarczyk, B = after Ganczarczyk and
Bachanek).

parallel, and show the tendency to activated sludge bulking at the very low
loadings and at the increased loadings. This feature is common for sewage
as well as for the Kraft pulp mill effluent. However, the curve for the fiber-
board mill effluent (Ganczarczyk and Bachanek, 1970) is displaced to the
right in comparison with the curve for sewage. It may show that the industrial wastewater has a higher tendency to cause activated sludge bulking at
low loadings than does sewage. This may be attributed to the high content
of carbohydrates. On the other hand, at increased loadings, this wastewater shows a lower tendency to cause activated sludge bulking. This may
be explained by its high pH value and alkalinity. The SVI data for sewage
given by Dart and Spur (1968) show the tendency of activated sludge to bulk
and deflocculate at loadings even lower than those of Eckenfelder (1967). In
the range of the higher loadings, the data of Ganczarczyk and Bachanek
(1970) show the optimal SVI value at loadings as high as 0.8 BOD/g.d., and
a comparatively moderate tendency to activated sludge bulking at lower and
higher loadings.

Another way to control the sludge volume index values is to extend the
retention of the activated sludge in the secondary clarifiers. This possible
method of achieving some decrease in SVI is well-known and often used in
the operational practice. The mechanism involved in these changes is not
quite clear. It is possible here that the prolonged anaerobiosis of the activated sludge is one of the causes, although the studies of Westgarth et al.
(1964) and Ford and Eckenfelder (1967) only partially confirm this hypothesis.
The other possibility is the physical thickening of the activated sludge being
retained in comparatively deep layers in secondary clarifiers. Two sets

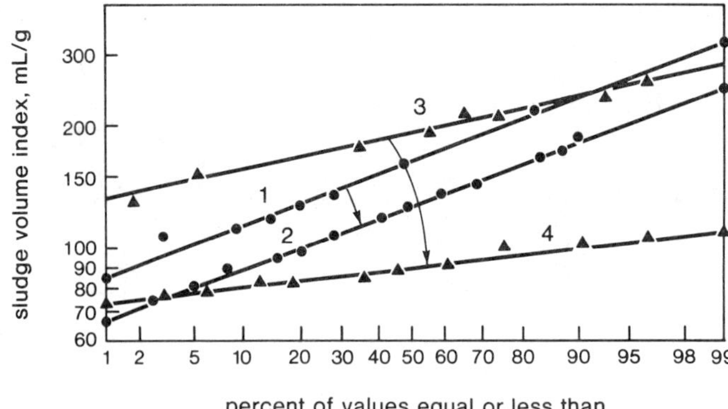

Figure 3-2.   Distribution in log-normal scale of sludge volume index data
from activated sludge treatment of the unbleached Kraft Pulp Mill effluent
treatment (1 and 2 and 3 and 4 = mixed liquor and return sludge for study
periods I and II respectively).

of measurements of SVI values of mixed liquor and recycled sludge in a
Kraft pulp mill effluent treatment plant are presented in Figure 3-2.   The
very distinct decreases of the SVI values for the recycled sludge may be
attributed to the oversized secondary clarifiers in this plant.

The level of concentration of the activated sludge in mixed liquor itself
has some influence on the SVI values (Isenberg et al., 1959, and Somers,
1968).

For chemical control of filamentous activated sludge bulking, an oxidative treatment by selective chlorination or by hydrogen peroxide additions may be used.   Filamentous growth develops a larger surface than
zoogleal growth and therefore it is more sensitive to such treatment.   It
was reported by Cole and Stamberg (1971) that additions of hydrogen peroxide destroyed the filamentous bacteria in activated sludge effectively decreasing the volume index.   It was believed that this was caused by an
oxygen "shock" effect or a peroxide-enzyme reaction.   At one municipal
installation, the addition of 20-40 mg/L hydrogen peroxide (H_2O_2) into
the activated sludge return pump decreased SVI values from 250-350 mL/g
to 50-60 mL/g after 24-48 hours' treatment.   Decreased SVI values re-
mained for more than two weeks following the hydrogen peroxide treatment.
In a later paper Cole et al. (1973) suggested the use of 20 to 200 mg/L of
H_2O_2 for a 24-hour period.   This required 1,668 lb/mgd (200 kg/1,000
cu.m), at a cost ranging from $55 to $85/mgd ($15 to $22/1,000 cu.m) according to 1971-1972 figures.

The other possible control method for activated sludge bulking is a
rapid change of some treatment parameters which can cause a required
shift in bacterial population dynamics.

The sludge volume index test does not measure all the physical properties of activated sludge. As an operational tool for in-plant control, the test is useful, but comparisons of sludge volume index measurements from various plants may not be very meaningful (Dick and Vesilind, 1969). Results of SVI tests cannot be used with certainty to predict settling behaviour in full-scale plants (see Chapter VI).

Chudoba et al. (1973) observed that the complete mixing systems encouraged excessive growth of some filamentous micro-organisms while the low degree of axial mixing conditions suppressed this growth. Therefore, the filamentous growth in the well mixed aeration tanks could be controlled to some extent by maintaining a substrate concentration gradient. The same group of researchers (1974) later confirmed the relationship between the sludge volume index and sludge load, but simultaneously observed different trends in this dependence at different flow patterns (complete mixing and plug-flow models). Sezgin et al. (1978) presented a unified theory of filamentous activated sludge bulking which comprises somewhat mechanistic concepts on activated sludge floc structure and behaviour. They postulated that the relative numbers of filamentous and zoogleal micro-organisms in the activated sludge floc were the most important factors in determining the physical properties of the activated sludge as related to settling and thickening. An excess of filamentous growth causes bulking of sludge, and not enough of it causes a formation of so-called "pin-point floc" which usually has a SVI below 70 mL/g, settles rapidly and leaves a turbid supernatant. A continuation of this work by Palm et al. (1980) showed that the dissolved oxygen concentration in aeration tanks has an important influence on the equilibrium between filamentous and zoogleal organisms in the activated sludge floc.

3.5 MIXED LIQUOR SUSPENDED SOLIDS

Mixed liquor suspended solids are composed of active microbial mass, nonactive microbial mass, non-biodegradable organics, and inorganic mass. In conventional activated sludge systems treating municipal wastewater, the active microbial mass usually represents only about 30 percent or less of mixed liquor suspended solids. In extended aeration activated sludge systems, the active microbial mass is usually less than 10 percent of this value. Weddle and Jenkins (1971) endeavoured to measure the unit viability of activated sludge in the form of the number of cells per mg of volatile suspended solids. They found that at a net growth rate of 0.3 day^{-1} (typical for a conventional activated sludge process) only 15 percent of the biomass could be considered active. This fraction, however, increased to 100 percent at growth rates as high as 3.0 day^{-1}. Therefore, an estimation of microbial density in aeration tanks by determining the mixed liquor suspended solids (MLSS) or, better still, the mixed liquor volatile suspended

solids (MLVSS) could only be imprecise because of the presence in this
material of variable concentrations of nonviable particulate matter.

The activity of the biomass in aeration tanks can be measured by oxygen
uptake rates, enumeration of viable cells by plate counting techniques, determination of dehydrogenase enzyme activity, estimation of adenosine triphosphate (ATP) and deoxyribonucleic acid (DNA) content. However, in
some studies it was indicated that only ATP content showed significant correlation to the organic carbon removal rate in a complex activated sludge
treatment system. An estimation of the biodegradable part of the biomass
as its active part was proposed by Barnard et al. (1972). Upadhyaye and
Eckenfelder (1975) obtained good correlations of this value for soluble wastewater with ATP, dehydrogenase activity and plate count. In a broad range
of sludge loadings (F/M) the variation was in the range of 3.39–3.56 x 10^{-9}
μg ATP per cell, and dehydrogenase activity varied from 20 to 25 μ moles of
TF per cell.

The level of mixed liquor suspended solids varies widely for various
modifications of the activated sludge process (see Chapter IV) and under
various modes of operation of the same modification. Generally, the higher
levels of MLSS call for higher oxygenation capacities in the system and require larger secondary clarifiers. However, the excess sludge resulting
from such treatment is somewhat smaller. Optimization analysis indicates
that the most economically attractive range of MLSS exists between 2,000
and 4,000 mg/L (CIRIA, 1975). To improve the operation of activated
sludge systems prone to sludge bulking, Kato and Sekikawa (1967) suggested
the installation of vertical plastic net panels in aeration tanks. As a substantial part of the activated sludge adheres to the net panels, the necessity
of recycling sludge from the final clarifiers diminishes as does the solids
loading of these units. The idea of this modification of the aeration tanks,
called "fixed activated sludge" tanks, is derived from the submerged contact
aeration system (Wilford and Conlon, 1957). Albert et al. (1972) studied
this approach for treatment of munition manufacturing wastewater.

3.6 DISSOLVED OXYGEN AND AERATION REQUIREMENTS

Many observations indicate that a concentration of 1 to 2 mg/L dissolved
oxygen in mixed liquor is sufficient for activated sludge treatment. However, some experiments by the British Coke Research Association (1971)
have shown that, for the treatment of concentrated coke plant effluents at
high organic loadings, the bacteria might be unable to utilize oxygen if a
higher dissolved oxygen tension was not maintained. Similar observations
have also been made in some studies on wastewater nitrification (Stenstrom
and Poduska, 1980), and, as was mentioned in p. 3.4, Palm et al. (1980)
found that dissolved oxygen concentration affects the equilibrium between
filamentous and zoogleal organisms in the biomass. It appears also that
the requirement for the minimum level of dissolved oxygen in aeration tanks

may be a function of micro-organisms dispersion. Therefore, it depends
on the mixing characteristics of the tanks (see p. 3.8) and the level of
mixed liquor suspended solids.

Commonly used figures for aeration requirements in the treatment of
sewage by means of a conventional activated sludge process are 1 cu. ft.
per gallon of primary effluent (7.5 cu. m/cu.m) or 1,000 cu. ft. per pound
of BOD applied (62 cu. m/kg). However, for safety reasons some design
standards suggest higher values (see Chapter IV). In the absence of molecular oxygen, the nitrate oxygen may be used to support the oxygen requirements of an activated sludge system (Schroeder and Busch, 1967). For the
treatment of some industrial wastewater containing a large concentration of
nitrate, utilization of this source of oxygen may be practical (Miyaji and
Kato, 1975).

Although gas transfer theory predicts that oxygenation is independent of
oxygen uptake rates, evaluation of several activated sludge plants shows a
relationship between uptake rates in the mixed liquor and the oxygen transfer capacity (Albertson and DiGregorio, 1975). This indicates that oxygen
requirements were not only limited by the oxygen rate as measured by
standard procedures and that an additional driving force for oxygen utilization existed. For consistently good treatment results it is also important
to maintain the required level of dissolved oxygen concentration in the
aeration tanks continuously. Roesler (1974) showed by processing data
from the Renton, Wash., Sewage Treatment Plant that automatic dissolved
oxygen control helped substantially to achieve an excellent operational performance (see also Chapter V, p. 5.8).

The dissolved oxygen level in aeration tanks affects the net sludge production and therefore some activated sludge modifications like the use of
commercial oxygen or deep-shaft aeration can produce less of the more
stabilized sludge (see Chapters II & VII). A common error in laboratory
experimentation on activated sludge treatment is to maintain the dissolved
oxygen levels markedly higher in model aeration tanks than those applied in
full-scale treatment. Such experiments, therefore, usually show a smaller
production of sludge, and, consequently, lead to an underestimation of this
treatment factor.

Pasveer (1955) found that a general relationship existed between the
oxygenation capacity of an aeration system (see Chapter V) and a BOD load
introduced to the aeration tank equipped with the given aerator(s). This observation was a basis to consider the aeration tank volume as a factor of a
secondary meaning only in a simplified technological design of some activated
sludge treatment systems (see Chapter IX).

3.7 SLUDGE AGE

In activated sludge treatment of wastewater, sludge age (S_A), or mean
solids (cells) residence time (SRT) in aeration tanks, is more characteristic

than wastewater aeration time (see p. 3.2). This former factor is closely
related to the reciprocal of net micro-organism specific growth rate and
can be presented as:

$$\frac{1}{S_A} = a\,\frac{F/M}{R} - b \tag{3-2}$$

where: F/M is food to micro-organisms ratio (see p. 3.3), "a" is cell
yield coefficient in mg cell per mg of substrate (see p. 2.8), "b" is the
micro-organisms endogenous decay coefficient in day^{-1}, and R is the
efficiency of BOD removal.

The value of sludge age may be determined by calculating the total
mass of micro-organisms in the activated sludge treatment system and
dividing it by the rate at which micro-organisms are intentionally and un-
intentionally wasted from the system. As it is fairly difficult to estimate
the amounts of activated sludge retained in secondary clarifiers, often
these amounts are ignored in the calculation of the sludge age, or taken
under account as the sludge in the clarifier sludge blanket only. The most
common approach to this calculation is based exclusively on the aeration
tank sludge inventory. For a completely mixed process operating at steady
state conditions, sludge age (S_A) is usually calculated by the relationship:

$$S_A = \frac{VZ}{Q_W Z_R + (Q-Q_W)Z_e} \tag{3-3}$$

in which: V = aeration basin volume; Q and Q_W = influent and wastage
volumetric flow rates respectively; and Z, Z_R, and Z_e = suspended solids
in aeration tank(s), recycle line, and final effluent respectively.

Example: for wastewater flow of 20,000 cu. m/day, aeration tank
volume of 5,000 cu. m, mixed liquor suspended solids level of 2,000
g/cu. m, excess activated sludge concentration of 12,000 g/cu. m, and the
treated wastewater residual suspended solids level of 20 g/cu. m, the
sludge age is:

$$S_A = \frac{5,000 \times 2,000}{200 \times 12,000 + (20,000 - 200)20} = 3.6 \text{ days}$$

Gould (1953) defined sludge age differently, as a ratio of sludge inventory
in aeration tanks to the daily load of suspended solids to these tanks. This
definition gained some popularity, but its technological meaning is very
limited.

In addition to the mean solids residence time, as expressed by the
equation (3-3), Lawrence and McCarthy (1970) developed a parameter desig-
nated as a design or operation safety factor and defined as a ratio between

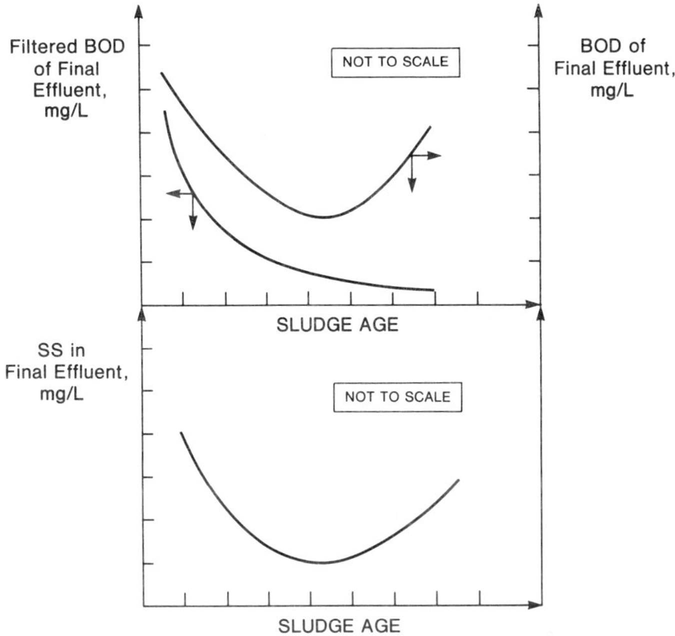

Figure 3-3. Influence of sludge age of final effluent quality (not to scale).

mean cell residence time and minimum cell residence time which the process can maintain without failing. Thus, high safety factor values are
associated with processes operating smoothly with little chance of upset
and small safety factors are associated with processes that are unstable
because of the greater probability that micro-organisms will be displaced
from the process at a rate greater than they can reproduce (see Chapter IX,
p. 9.2).

Maintaining a constant mean solids residence time in aeration tanks is
advocated as the best method for controlling the activated sludge system
(Garrett, 1958; Jenkins and Garrison, 1968; Walker, 1971; and Burchett
and Tchobanoglous, 1974). If excess sludge is wasted directly from aeration tanks, as a portion of mixed liquor, the wasting rate is independent of
the mixed liquor and return sludge concentrations, and is equal only to the
total volume of the aeration tanks divided by the desired cell retention time.
This is called the hydraulic control of the activated sludge process and can
be easily automated (see Chapter X, p. 10.3).

Bisogni and Lawrence (1971) studied a case of activated sludge treatment of a soluble substrate, and showed a relationship between the sludge
age (S_A) and the resulting sludge volume index (SVI) and zone settling velocity (ZSV). It was concluded that sludge settling properties are best at

higher values of the sludge age, where the sludge volume index is minimal and zone settling velocity is maximum. To some extent, mineral nutrient requirements may also be a function of the maintained sludge age, as higher values of this factor facilitate the internal recycle of nutrients. In another study Saunders and Dick (1981) determined the influence of the sludge age on the characteristics of organic matter in sewage treatment effluent. With the sludge age increased from less than 1 day to almost 20 days, significant reductions in filtrable effluent COD were achieved, and decreases in molecular weight of colloidal fraction of the studied substances were observed.

Based on an analysis of numerous cases of activated sludge process applications, it seems possible to determine optimal sludge age values to minimize the residual BOD values and residual suspended solids in a given type of treated wastewater (Fig. 3-3).

3.8 EFFECTS OF MIXING AND TURBULENCE

Mixing and turbulence in activated sludge aeration tanks may be considered in macro- and micro-scale. Macro-scale mixing promotes uniformity in the system. Ideal macro-scale mixing can be described by constant concentration of suspended solids and dissolved and/or colloidal organics throughout the whole tank volume. On the other hand micro-mixing is pertinent to the molecular diffusion mechanism of transport phenomena (see Chapter II, p. 2.6), influencing directly the film thicknesses across which migrate the molecules of biochemical reaction substrates and products.

The mixing analysis of aeration tanks is characterized by conditions between two extreme situations: tubular reactor (plug or piston flow) and completely mixed reactor (Fig. 3-4). Plug flow situations are often assumed to be analogous to those existing in the batch type reactors with an elapse of the observation time. On the other hand, in aeration tanks which are relatively long and narrow, concentration gradients of characteristic quality parameters of the treated wastewater exist at the same time at different observation points. Mixing in aeration tanks was the subject of extensive studies by Grieves et al. (1964), Milbury et al. (1965), Murphy and Timpany (1967), Tucek and Chudoba (1969), Murphy and Boyko (1970), Toerber et al. (1974), etc.

Mixing and turbulence in aeration tanks is accomplished mostly by the movement of air bubbles produced by the introduction of compressed air through various media, or by operation of various mechanical aerators (see Chapter V). It appears that high turbulence levels in aeration tanks may adversely affect reflocculation abilities of activated sludge. The role of the mixed liquor flow is negligible and application of additional mixer is justified only by the type of aeration system applied. Air bubble rising velocity is a function of bubble diameter and surface tension and density of mixed liquor. However, a "plateau" of bubble rising velocity exists over

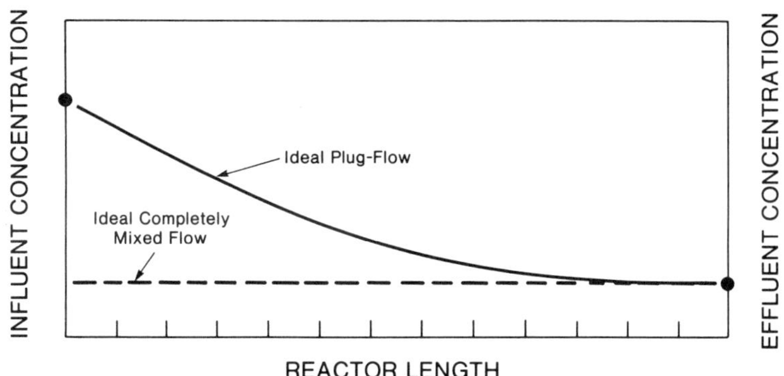

Figure 3-4. Plug flow and completely mixed flow activated sludge reactors.

the range of diameter from 2 mm to 20 mm. The cause of this phenomenon lies in the change of bubble shape from spherical in laminar flow to a cap-form in turbulent regime. The mixing characteristics of mechanical aerators is their individual feature.

Residence time distributions can be used for characterization of mixing degree in an aeration tank, but they do not relate to its homogeneity. The residence time distribution specifies the fraction of mass or increments of flow which remain in a tank for less than time t. The time derivative of this value is the residence time frequency distribution and refers to the fraction of mass or flow increments which remain in a tank between time t and time t + dt. Another general measure of degree of mixing in aeration tanks may be an estimation of energy input into the given tank.

Observations of Chudoba et al. (1973) seemed to indicate that the complete mixing systems tend to lead to excessive growth of the filamentous micro-organisms. Conversely, the aeration systems with a low degree of axial mixing may suppress the growth of filaments. Therefore, it appears that by maintaining the concentration gradient of the substrate along the aeration tank it is possible to control, at least to some extent, the activated sludge bulking.

3.9 EFFECTS OF WASTEWATER TEMPERATURE

The influence of temperature on activated sludge treatment is complex. An increase of temperature decreases the wastewater viscosity and its surface tension. Therefore, it improves mixing phenomena (see p. 3.7) and molecular diffusion of substrates and products of biochemical reactions. Such an improvement in molecular diffusion of oxygen and of turbulent transport

usually overcomes slightly a decrease in oxygenation capacity caused by a
decrease in oxygen deficit due to the temperature increase (see Chapter V).
Some increase in wastewater temperature often improves the performance
of secondary settlers also.

Individual species of micro-organisms show specific optimum temperatures for growth and substrate utilization. This leads to their classification
into psychrophiles, mesophiles and thermophiles. The specific temperature
ranges for these groups of organisms are only loosely defined. The temperature dependence of micro-organism growth may be explained by temperature
dependent enzymatic transformations, specific to particular species. In
addition, it must be pointed out that not only the environmental temperature
but also a temperature change affects the microbial physiology.

The temperature dependence of the rate constant (k) for a well defined
chemical reaction is given by the van't Hoff-Arrhenius equation:

$$\frac{d(\ln k)}{dT} = \frac{E}{RT^2} \tag{3-4}$$

where: T = temperature in $°K$; R = ideal gas constant (199 cal. per $°K$
per mole); and E = activation energy in cal. per mole. Integration of this
equation between the limits T_1 and T_2 gives:

$$\ln \frac{K_2}{K_1} = \frac{E(T_2 - T_1)}{R\,T_1\,T_2} \tag{3-5}$$

Assuming that E/RT_1T_2 is constant for practical situations in wastewater
treatment, leads to the relationship:

$$\ln \frac{K_2}{K_1} = C\,(T_2 - T_1) \quad \text{or} \tag{3-6}$$

$$\frac{K_2}{K_1} = e^{C(T_2 - T_1)} \tag{3-7}$$

Furthermore, replacing e^C with a temperature coefficient $\textcircled{H}$ gives:

$$K_2 = K_1\,\theta^{(t_2 - t_1)} \tag{3-8}$$

In this expression $°K$ can also be replaced by $°C$. A similar relationship
often employed to describe the influence of mixed liquor temperature on
activated sludge process kinetics is:

$$k_t = k_{20°C} \, \theta^{(t-20)} \tag{3-9}$$

Although an application of the van't Hoff-Arrhenius equation to complex phenomena of biological oxygen utilization and growth does not have any theoretical justification, the Eq. 3-9 was empirically proven under some experimental conditions. Wuhrman (1954) found that for activated sludge loadings lower than 0.5 lb (g) BOD per day lb (g) MLSS, θ was equal to 1.0. However, for higher loadings values of θ ranged from 1.0 to 1.04. The same author in more recent pilot-plant studies (Wuhrman, 1964) confirmed the same temperature dependence at higher loadings. The lack of sensitivity of the activated sludge process to temperature at low loadings was explained by Eckenfelder (1970) on the basis that at higher temperatures, a higher bacterial respiration rate keeps only part of the bacterial floc in an active aerobic condition; at lower temperatures, a larger number of bacteria is aerobically active at the same or similar rate of oxygen diffusion into the floc, giving practically the same treatment efficiency. However, at higher loadings, bacterial flocs are usually smaller, and therefore, the number of active bacteria cannot be self-regulated this way. According to Eckenfelder (1975) the θ values are significant only for soluble organic substances and vary from 1.055 to 1.10. For predominantly colloidal wastewater, like sewage, temperature dependence of the process may be negligible. Viehl (1940) in his studies of temperature influence on activated sludge treatment found that there were some performance differences in laboratory scale experiments, but none in full-scale observations. His explanation was that there were some limitations of bacteria adjustment to temperature changes if only a smaller sample of bacteria was involved.

Temperature coefficients for the activated sludge process cover the influence of temperature on: (a) diffusion phenomena, (b) sorption on bacterial mass, (c) exoenzymatic hydrolysis, (d) bacterial growth, and (e) bacterial endogenous respiration. Simultaneously, temperature adaptation capabilities of various bacteria vary drastically. Therefore, it is not surprising that these values are not constant but vary widely, as demonstrated by Henry (1971), Friedman and Schroeder (1972) and others. It was observed in several situations that the Arrhenius equation may not apply to mixed cultures, and that sludge recycle could offset potential temperature effects. Novak (1974) found that the temperature response of biological processes depended also on the substrate concentration in the system (see p. 3.10).

Aerobic digestion of sludge, and nitrification and denitrification processes are much more sensitive to temperature than biological oxidation of organic substances in the activated sludge process. Moreover, mixed liquor temperatures may also affect the percentage of nonsettleable suspended solids in the treated wastewater. In several plants it was observed that average suspended solids content in the effluent was higher in winter than in summer, and this could not be explained only by a different functioning of the secondary clarifiers.

Effects of shock temperature changes on activated sludge systems were investigated by Carter and Barry (1975). Both stoichiometric and rate inhibition effects were observed. In some cases, activated sludge treatment has been adversely affected by temperature increases. For example, in a pulp mill effluent treatment plant in Covington, Va. it was found (Burns, 1963) that an increase of wastewater temperature over 100° F (38° C) caused a substantial decrease in the treatment efficiency. Therefore, to maintain high BOD removal, cooling of wastewater prior to the treatment was introduced during the summer months (see Chapter VIII).

The temperature of wastewater is usually higher than that of the water supply. In most of the United States and the southern part of Canada, mean sewage temperatures vary from about 50 to 70° F (10-21° C), with the most common value of 60° F (15.6° C). Aerobic microbial transformations of organic substances can generate heat. This phenomenon is usually unimportant in the activated sludge wastewater treatment process because of relatively low wastewater concentrations and wastewater cooling effects of evaporation associated with intensive aeration. However, with treatment of highly concentrated wastewater (Poepel and Ohnmacht, 1972), or the aerobic digestion of organic sludges, especially with the use of commercial oxygen for aeration, which minimizes evaporation effects (Smith et al., 1975), the process heat production may be sufficient to support treatment in a thermophilic range of temperatures (see also Chapter VII).

There are conflicting reports about how temperature affects activated sludge treatment in large scale plants. Bloodgood (1944) concluded that this factor played an important role in determining the Indianapolis plant capacity; however, Keefer (1962) observed that BOD decreases were only slightly improved at higher temperatures in the Baltimore Back River Plant. According to the statistical studies of Lin and Heinke (1975), temperature proved to be the second most important factor influencing the effectiveness of the activated sludge process. The analysis of data from 13 plants for 25 years of operation revealed an adverse effect of low temperature on BOD and suspended solids removal. The effect of temperature on primary sedimentation was negligible.

3.10 EFFECTS OF WASTEWATER CONCENTRATION

The strength of wastewater, which is a common substrate for the activated sludge process, usually varies from about 100 to 300 mg/L BOD. However, when a very diluted or very concentrated wastewater has to be treated, several technological problems can arise. In the case of diluted wastewater, the relatively high level of residual organics in the treated wastewater may drastically decrease the activated sludge process efficiency. In the case of concentrated wastewater, to achieve the required treatment, even dilution of the wastewater is sometimes suggested (see Chapter VIII). The current trend for water use in industry is the limitation of quantity consumed and the application of closed water circuits. This leads to the receiving of smaller volumes of more concentrated wastewater. The treatment of this wastewater to meet the specific requirements of receiving waters is a relatively new task for wastewater technology. The influence of wastewater concentration on activated sludge process performance has been the subject of comparatively few studies carried out in a relatively narrow range of concentration values (Wilson, 1962; Braun, 1966; and Graham, 1968), although this problem is not a new one in the alcoholic fermentation and yeast propagation industry, Ganczarczyk and Bachanek (1970) studied activated sludge treatment of fibreboard mill effluents at concentrations ranging from 7,300 to 42,000 mg/L COD. This wastewater contained the complicated mixture of dissolved and colloidal organic substances. These were primarily colloidal carbohydrates, and to a lesser extent furfurol, lower aliphatic acids, alcohols, aldehydes and ketones. In the 11 series of experiments, loading parameters in the range from 0.4 to 4.25 g COD/g MLSS per day, and aeration periods from 2.9 to 123 hours were investigated. Even with extreme conditions, a failure of the process was not observed, although only limited treatment results were obtained. All the results of these experiments could be covered by an equation for a retardant reaction in which the rate constant and the retardation coefficient surprisingly proved to be functions of the initial concentration of the wastewater studied. It was also observed that for the highly concentrated wastewater it was impossible to reduce the effluent concentration in one-stage process below some specific level related to the influent concentration. This phenomenon was explained as a consequence of an accumulation of inert and low-rate biodegradable materials in the mixed liquor. For other concentrated wastewaters, containing only easily biodegradable substances, the above observation has not been repeated.

According to the generally accepted models of the activated sludge process, the effluent substrate concentration is independent of the influent substrate concentration. Grady et al. (1972) confirmed this when working with pure bacterial cultures and a single metabolic substrate. These findings were not confirmed, however, for mixed cultures or with a composite determination of organic substrate (such as BOD, TOC, COD, etc.) which show the effluent concentrations increasing at the influent higher concentrations by about 1 percent of the latter concentration (Grady and Williamson, 1975). This theoretical study showed that in an activated sludge system the mean solids residence time (sludge age) must be increased as the influent concentration is increased in order to maintain the same effluent quality.

Adams et al. (1975) modified a general activated sludge treatment kinetic equation (see Chapter II) for highly concentrated wastewaters. There are also indications that a high total dissolved solids concentration in some wastewaters affects bacterial flocculation and causes an increase in suspended solids levels in the treated effluents (see Chapter VIII, p. 8.1).

REFERENCES

Adams, C. E., et al.: "A Kinetic Model for Design of Completely Mixed Activated Sludge Treating Variable-Strength Industrial Wastewater," Water Research (Brit.), 9, 37 (1975).

Albert, R. C., Hoehn, R. C., and Randall, C. W.: "Treatment of a Munitions-Manufacturing Waste by the Fixed Activated Sludge Process," paper presented at the 27th Purdue Industrial Waste Conference, Lafayette, Indiana, 1972.

Albertson, O. E., and DiGregorio, D.: "Biologically Mediated Inconsistencies in Aeration Equipment Performance," Jour. Water Poll. Control Fed., 47, 976 (1975).

Ardern, E., and Lockett, W. T.: "Experiments on the Oxidation of Sewage Without Filters," Jour. Soc. Chem. Ind., 33, 521 (1914).

Batow, E. and Mohlman, F. W.: "Sewage Treatment Experiments with Aeration and Activated Sludge," Eng. News, 73, 647 (1915).

Barnard, J. L., et al.: "Design Optimization for Activated Sludge and Extended Aeration Plants," paper presented at the 6th International Conference on Water Pollution Research, Jerusalem, Israel, 1972.

Bisogni, J. J., and Lawrence, A. W.: "Relationships Between Biological Solids Retention Time and Settling Characteristics of Activated Sludge," Water Research (Brit.) 5, 753 (1971).

Bloodgood, D. E.: "The Effect of Temperature and Organic Loading Upon Activated Sludge Plant Operation," Sewage Works Jour., 16, 913 (1944).

Braun, H.: "Der Einfluss der Nahrstoffkonzentration auf die Reaktionskinetik des Belebtschlammverfahvens," Karlsruher Berichte zur Ingenieurbiologie No. 1, Karsruhe 1966.

British Coke Research Assoc.: "A Study of the Oxygen Requirement in the Treatment of Liquid Effluents by the Activated-Sludge Process," Coke Research Report No. 66, June 1971.

Burchett, M. E., and Tchobanoglous, G.: "Facilities for Controlling the Activated Sludge Process by Mean Cell Residence Time," Jour. Water Poll. Control Fed., 46, 973 (1974).

Burns, O. B.: "Recent Improvement Studies at the West Virginia Pulp and Paper Waste Treatment Plant in Covington, Va.", National Council for Steam Improvement, August, 1963.

Carter, J. L., and Barry, W. F.: "Effects of Shock Temperature on Biological Systems," Proc. Amer. Soc. Civ. Engrs., Jour. Environ. Eng. Div., 101, 229 (1975).

Chudoba, J., et al.: "Control of Activated Sludge Filamentous Bulking.
I. Effect of the Hydraulic Regime or Degree of Mixing in an Aeration
Tank," Water Research (Brit.), 7, 1163 (1973).

Chudoba, J., et al.: "Control of Activated Sludge Filamentous Bulking.
II. Selection of Microorganisms by Means of a Selector," Water Research
(Brit.), 7, 1389 (1973).

Chudoba, J., et al.: "Control of Activated Sludge Filamentous Bulking.
III. Effects of Sludge Loading," Water Research (Brit.), 8, 231 (1974).

Cole, C. A., and Stamberg, J. B.: "Hydrogen Peroxide Cures Filamentous Bulking Activated Sludge," paper presented at the 5th Mid-Atlantic
Industrial Waste Conference, November 1971.

Cole, C. A., et al.: "H_2O_2 Cures Filamentous Growth in Activated Sludge."
Jour. Water Poll. Control. Fed., 45, 829 (1973).

CIRIA — Construction Industry Research and Information Association (G. B.):
"Cost-Effective Sewage Treatment. An Assessment of the Prototype Model,"
Report 54 (1975).

Dart, M. C., and Spurr, T.: "Treatment of Domestic Sewage by the Contact Stabilization Process," Water and Waste Treat. Jour., 12, 5/6, 12
(1968).

Dick, R. J., and Vesilind, P. A.: "The Sludge Volume Index — What Is It?,"
Jour. Water Poll. Control Fed., 41, 1285 (1969).

DiToro, D. M.: "Simplified Equalization Basin Design," Proc. 29th Ind.
Waste Conference, Purdue Univ., Ext. Ser., 135 (1974).

DiToro, D. M.: "Statistical Design of Equalization Basins," Proc. Amer.
Soc. Civ. Engrs,, Jour. Env. Eng. Div., 100, EE6, Paper 11782, (1975).

Duggan, J. B., and Cleasby, J. L.: "Effect of Variable Loading on Oxygen
Uptake," Jour. Water Poll. Control Fed., 48, 540 (1976).

Eckenfelder, W. W., Jr.: "Comparative Biological Waste Treatment
Design," Proc. Amer. Soc. Civ. Engrs., Jour. San. Eng. Div., 93, SA6,
157 (1967).

Eckenfelder, W. W., Jr.: "Water Quality Engineering for Practicing
Engineers," Barnes and Noble, 1970.

Eckenfelder, W. W., Jr.: "Wastewater Treatment Design," Water & Sewage Works, 122, 6, 62 (1975).

Eckhoff, D. W., and Jenkins, D.: "Transient Loading Effects in the Activated Sludge Process," in "Advances in Water Pollution Research," Proc.
3rd Int. Conf. on Water Pollution Research, Munich 1966.

Foess, G. W., et al.: "Evaluation of In-line and Side-line Flow Equalization Systems," paper presented at the 48th Conference of Water Poll. Control Fed., Miami Beach, Fla., 1975.

Ford, D. L., and Eckenfelder, W. W., Jr.: "Effect of Process Variables on Sludge Floc Formation and Settling Characteristics," Jour. Water Poll. Control Fed., 39, 1850 (1967).

Friedman, A. A., and Schroeder, E. D.: "Temperature Effects on Growth and Yield of Activated Sludge," Jour. Water Poll. Control Fed., 44, 1433 (1972).

Ganczarczyk, J., and Bachanek, S.: "The Influence of the Wastewater Concentration on the Activated Sludge Process Efficiency," paper presented at the 5th International Conference on Water Pollution Research, San Francisco, Calif., 1970.

Ganczarczyk, J.: "Variation in the Activated Sludge Volume Index," Water Research (Brit.), 4, 69 (1970).

Ganetelli, E. J., and Heukelekian, H.: "The Influence of Loading and Chemical Composition of Substrate on the Performance of Activated Sludge," Jour. Water Poll. Control Fed., 36, 643 (1964).

Garret, M. T., Jr.: "Hydraulic Control of Activated Sludge Growth Rate," Sewage and Ind. Wastes, 30, 253 (1958).

Gould, H. R.: "Sewage Aeration Practice in New York City," Proc. Am. Soc. Civ. Eng., Jour. San. Eng. Div., 79, 307 (1953).

Grady, C. P. L., et al.: "Effects of Growth Rate and Influent Substrate Concentration on Effluent Quality from Chemostat Containing Bacteria in Pure and Mixed Culture," Biotech. Bioeng., 14, 391 (1972).

Grady, C. P. L., and Williamson, D. R.: "Effect of Influent Substrate Concentration on the Kinetics of Natural Microbial Populations in Continuous Culture," Water Research (Brit.), 9, 171 (1975).

Graham, P. W.: "Kinetic Aspects of the Treatment of Phenolic Wastes," paper presented at the 4th International Conference on Water Poll. Research, Prague, Czechoslovakia, 1969.

Grieves, R. B., et al.: "A Mixing Model for Activated Sludge," Jour. Water Poll. Control Fed., 36, 619 (1964).

Haseltine, T. R.: "A Rational Approach to the Design of Activated Sludge Plants," Water & Sewage Works, 102, 487 (1955).

Henry, J. G.: "Microbial Response in Low Temperature Waste Treatment," Ph.D. Thesis, University of Toronto, Toronto, 1971.

Heukelekian, H., and Weisberg, E.: "Bound Water and Activated Sludge Bulking," Sewage Ind. Wastes, 28, 558 (1956).

Isenberg, E., and Heukelekian, H.: "Sludge Volume Index," Water Sewage Wks. 106, 525 (1959).

Jones, P. H.: "The Effect of Nitrogen and Phosphorus Compounds on One of the Microorganisms Responsible for Sludge Bulking," paper presented at the 20th Purdue Ind. Waste Conference, Purdue University, Lafayette, Ind., 1965.

Jenkins, D., and Garrison, W. E.: "Control of Activated Sludge by Mean Cell Residence Time," Jour. Water Poll. Control Fed., 40, 1905 (1968).

Kato, K., and Sekikawa, Y.: "Fixed Activated Sludge Process for Industrial Waste Treatment," Proc. of the 22nd Ind. Waste Conf., Purdue University, Lafayette, Indiana, Eng. Ext. Series No. 129, 926 (1967).

Keefer, C. E.: "Temperature and Efficiency of the Activated Sludge Process," Jour. Water Poll. Control Fed., 34, 1186 (1962).

Knopp, P. V., et al.: "Transients in the Activated Sludge Process," Proc. 23rd Ind. Waste Conf., Purdue Univ., Lafayette, Inc., 1968.

Krishnan, P., and Gaudy, A. F., Jr.: "Response of Activated Sludge to Quantitative Shock Loading," Jour. Water Poll. Control Fed., 48, 906 (1976).

Kuslikis, B. P.: "An Evaluation of Two New High Rate Biological Wastewater Treatment Technologies," Master's Thesis, University of Toronto, Toronto, 1978.

Lawrence, A. W., and McCarty, P. L.: "Unified Basis for Biological Treatment Design and Operation," Proc. Am. Soc. Civil Eng., Jour. Sanit. Eng. Div., 96, SA3, 757 (1970).

Leiser, C. P.: "Computer Management of a Combined Sewer System," USEPA Report No. EPA-670/2-74-022, National Environmental Research Centre, Cincinnati, Ohio, 1974.

Lin, Kwan-Chow, and Heinke, G. W.: "Plant Data Analysis of Temperature Significance in the Activated Sludge Process," Jour. Water Poll. Control Fed., 49, 286 (1977).

Logan, R. P., and Budd, W. E.: "Effect of BOD Loadings on Activated Sludge Plant Operation," in "Biological Treatment of Sewage and Industrial Wastes," Vol. 1, p. 271, Reinhold Publ. Co., New York, 1956.

Milbury, W. F., et al.: "Compartmentalization of Aeration Tanks," Proc. Amer. Soc. Civil Engr., Jour. San. Eng. Div., 91, SA 3, 45 (1965).

Miyaji, Y., and Kato, K.: "Biological Treatment of Industrial Wastes Water by Using Nitrate as an Oxygen Source," Water Research (Brit.), 9, 95 (1975).

Mohlman, F. W.: "The Sludge Index," Sewage Works Jour., 6, 119 (1934).

Murphy, K. L., and Boyko, B. I.: "Longitudinal Mixing in Spiral Flow
Aeration Tanks," Proc. Amer. Soc. Civil Engr., Jour. San. Eng. Div.,
96, SA 2, 211 (1970).

Murphy, K. L., and Timpany, P. L.: "Design and Analysis of Mixing for
an Aeration Tank," Proc. Amer. Soc. Civil Engr., Jour. San. Eng. Div.,
93, SA5, 1 (1967).

Novak, J. T.: "Temperature — Substrate Interactions in Biological Treatment," Jour. Water Poll. Control Fed., 46, 1984 (1974).

Novotny, V., and Englande, A. J. Jr.: "Equalization Design Techniques
for Conservative Substances," Water Research (Brit.), 8, 325 (1974).

Novotny, V., and Stein, R. M.: "Equalization of Time Variable Waste
Loads," Proc. Amer. Soc. Civ. Engrs., Jour. Env. Eng. Div. ASCE,
102, EE3, 613 (1976).

Palm, J. C., et al.: "The Relationship Between Organic Loading, Dissolved
Oxygen Concentration, and Sludge Settleability in the Completely-Mixed
Activated Sludge Process," Jour. Water Poll. Control Fed., 52, 2485
(1980).

Pasveer, A.: "Research on Activated Sludge. V. Rate of Biochemical
Oxidation," Sewage and Ind. Wastes, 27, 783 (1955).

Pipes, O. W.: "Types of Activated Sludge Which Separate Poorly," Jour.
Water Poll. Control Fed., 41, 714 (1969).

Poepel, F., and Ohnmacht, C.: "Thermophilic Bacterial Oxidation of
Highly Concentrated Substrates," Water Research (Brit.), 6, 807 (1972).

Roesler, J. F.: "Plant Performance Using Dissolved Oxygen Control,"
Proc. Amer. Soc. Civ. Engrs., Jour. Envir. Eng. Div., 100, EE5, 1069
(1974).

Ruchhoft, C. C., and Kochmar, J. P.: "The Role of Spaerotilus Natans in
Activated Sludge Bulking," Sewage Works Jour., 13, 3 (1941).

Saunders, F. M., and Dick, R. I.: "Effect of Mean-Cell Residence Time
on Organic Composition of Activated Sludge Effluents," Jour. Water Poll.
Control Fed., 53, 201 (1981).

Saleh, M. M., and Gaudy, A. F. Jr.: "Shock Load Response of Activated
Sludge with Constant Recycle Sludge Concentration," Jour. Water Poll.
Control. Fed., 50, 764 (1978).

Schroeder, E. D., and Busch, A. W.: "The Role of Nitrate Nitrogen in
Bio-Oxidation," Proc. 22nd Ind. Waste Conf., Purdue Univ., Lafayette,
Ind., 1967.

Selma, M. W., and Schroeder, E. D.: "Response of Activated Sludge Process to Organic Transients — Kinetics," Jour. Water Poll. Control Fed., 50, 944 (1978).

Sezgin, M., et al.: "A Unified Theory of Filamentous Activated Sludge Bulking," Jour. Water Poll. Control Fed., 50, 362 (1978).

Smitt, J.: "Bulking of Activated Sludge. Causative Organisms," Sewage Works Jour., 6, 1041 (1934).

Smith, J. E. Jr., Young, K. W., and Dean, R. B.: "Biological Oxidation and Disinfection of Sludge," Water Research (Brit.), 9, 17 (1975).

Smith, R., et al.: "Design and Simulation of Equalization Basins," U.S. EPA, National Environmental Research and Development Centre, Report EPA-670/7-73-046 (1973).

Somers, J. A.: "The Relation Between Sludge Volume Index and Sludge Content in the Activated Sludge Process," Water Research (Brit.), 2, 563 (1968).

Speece, R. E., and LaGrega, M.: "Flow Equalization by Use of Aeration Tank Volume," Jour. Water Poll. Control Fed., 48, 2599 (1976).

Stenstrom, M. K., and Poduska, R. A.: "The Effect of Dissolved Oxygen Concentration on Nitrification," Water Research (Brit.), 14, 643 (1980).

Toerber, E. D., et al.: "Comparison of Completely Mixed and Plug Flow Biological Systems," Jour. Water Poll. Control Fed., 46, 1995 (1974).

Tucek, F., and Chudoba, J.: "Purification Efficiency in Aeration Tanks with Complete Mixing and Piston Flow," Water Res. (Brit.), 3, 559 (1969).

Upadhyaya, A. K., and Eckenfelder, W. W., Jr.: "Biodegradable Fractions as an Activity Parameter of Activated Sludge," Water Research (Brit.), 9, 691 (1975).

Viehl, K.: "Der Einfluss der Temperatur auf die Selbstreinigung des Wassers . . .," Zeit. f. Hygiene u. Infectionskrank., 122, 81 (1940).

Walker, L. F.: "Hydraulically Controlling Solids Retention Time in the Activated Sludge Process," Jour. Water Poll. Contr. Fed., 43, 30 (1971).

Wallace, A. T.: "Analysis of Equalization Basins," Proc. Amer. Soc. Civ. Eng., Jour. San. Eng. Div., 94, SA6, 1161 (1968).

Weddle, C. L., and Jenkins, D.: "The Viability and Activity of Activated Sludge," Water Research 5, 621 (1971).

Westgarth, W. C., Sulzer, F. T., and Okun, D. A.: "Anaerobiosis in the Activated Sludge Process," paper presented at the 2nd International Conference on Water Poll. Research, Tokyo, Japan, 1964.

Wilford, J., and Conlon, T. P.: "Contact Aeration Sewage Treatment Plants in New Jersey," Sew. and Ind. Wastes, 29, 845 (1957).

Wilson, J. S.: "Concentration Effects in the Biological Oxidation of Trade Wastes," paper presented at the 1st International Conf. on Water Poll. Research, London, England, 1962.

Wuhrmann, K.: "High Rate Activated Sludge Treatment and its Relation to Stream Sanitation," Jour. Water Poll. Control Fed., 26, 1 (1954).

Wuhrmann, K.: "Hauptwirkungen und Wechselwirkungen einiger Betriebsparameter im Belebtschlammsystem," paper presented at the EAWAG Conference, Zurich, Switzerland, April 1964.

Modification of Activated Sludge Process

4.1 BASIC PROCESS MODIFICATIONS

Particular modifications of the activated sludge process can be undertaken within different ranges of the process parameters (Chapter III), and can produce effluents of different quality. The basis for the selection of a required process modification is a determination of treatment objectives, characteristic of wastewater, and the size of the treatment plant. However, local examples, and a specific experience of the consulting engineers or water authorities may also have some influence on this decision.

In general, the low-rate modifications operate at a low food to micro-organism ratio (F/M) and at a high sludge age value (S_A). They produce effluents of higher quality (often nitrified), and produce less of the excess sludge. They also show more "resistance" to shock loads. Conversely, the high-rate modifications operate at a high food to micro-organisms ratio, and at a low sludge age value. They produce effluents of an inferior quality and more of the excess sludge. Their resistance to shock loads is less and they may be less effective for toxicity and colour removal from some industrial effluents.

Mechanical pre-treatment of wastewater prior to activated sludge treatment is optional from a technological point of view, but it may be an important factor economically.

Table 4-1 shows the ideal ranges of volumetric and sludge loadings (Chapter III, p. 3.3), mixed liquor suspended solids levels (Chapter III, p. 3.5), sludge age (Chapter III, p. 3.7), oxygenation capacity to load ratios (Chapter III, p. 3.6), and the resulting wastewater BOD decrease, which is characteristic of municipal sewage treatment according to the basic modifications of the activated sludge process. In practice, the demarcation lines among some modifications of the process are not well drawn (see Tables 4-2, 4-3, 4-4, 4-6, 4-11, and 4-12).

TABLE 4-1
Basic Modifications of Activated Sludge Process

Modification	Volumetric Loadings in BOD lb/1000 cu. ft. per day (in BOD g/cu. m per day)	Sludge Loadings in BOD g/ g MLSS per day	MLSS Level in mg/L	Sludge Age in days	OC/L Ratio	BOD Decrease in %
Extended Aeration						
(a) oxidation ditches	5.5–13 (100–200)	0.02–0.10	3500–5000	60–90	>2	>90
(b) package plants	up to 30 (up to 480)	0.12–0.50	3500–5000	up to 40	>2	75–90
Conventional	30–45 (500–700)	0.10–0.60	1500–3000	3–4	1.5	90
High Efficiency	300–1550 (5000–25000)	up to 2.7	3000–8000	–	1.5	70–90
Partial Treatment	100–200 (1700–3200)	up to 4.6	300–700	0.2–0.5	<1.0	50–70

4.2 EXTENDED AERATION SYSTEMS

Extended aeration is a low load modification of the activated sludge process, characterized by long retention times of mixed liquor in aeration tanks and high content of activated sludge suspended solids in mixed liquor. This modification leads to relatively low amounts of well-stabilized excess biomass. Therefore, it is sometimes called "total oxidation" of wastewater. This modification has excellent equalization properties in case of shock discharges of wastewater, and is often applied in package-type plants (see Chapter IX, p. 9.6) and in oxidation ditches (see Chapter V, p. 5.7).

Package-type extended aeration facilities usually work without primary clarification and are designed for an aeration tank detention time of 24 hours and 4 hours retention in secondary clarifiers. A small fraction of the settled sludge in such units is often subjected to aerobic digestion with infrequent (e.g., every six months or every year) disposal of digested sludge. A broad selection of pre-fabricated package plants is available from various producers.

The "Ten States Standards" (Great Lakes–Upper Mississippi River Board of State Sanitary Engineers, 1973) suggest, for extended aeration systems, 24 hours aeration detention, aeration tank volume loading of 12.5 lb BOD per 1,000 cu. ft. per day (200 g BOD per cu. m. per day) and substrate loadings between 0.05 and 0.1 lb BOD per lb MLSS per day (g BOD per g MLSS per day). Simultaneously, these Standards call for minimum air requirements of 2,000 cu. ft. per lb BOD aeration tank load (124.6 cu. m per kg BOD), and an average sludge recycle of 100 percent (50 percent minimum and 200 percent maximum).

The extended aeration modification of the activated sludge process often employs oxidation ditches as biological reactors (Fig. 4-1). This reactor was originally developed for this type of activated sludge process (Pasveer, 1957 and 1958; Baars, 1962), but can be also used for some other process modifications. Volumetric and sludge loadings applied in oxidation ditch treatment of small flows of wastewater are usually lower than the parameter values selected for the "package aeration" plans (see Table 4-1).

4.3 CONVENTIONAL SYSTEMS

Relatively early in the development of the activated sludge process, it was empirically established that, to achieve 90 percent removal (or more) of municipal wastewater BOD, aeration times of at least 6 to 8 hours were required. Volumetric loadings of 20 to 50 lb BOD/day/1,000 cu. ft. (320 to 800 kg/day/1,000 cu. m) were to be applied, as well as a food to microorganisms ratio of around 0.35. As this treatment approach and efficiency is very popular, the respective process modification is commonly called "conventional," and other modifications are usually compared with it. This

Figure 4-1. Double oxydation ditch in Southampton, Ontario (photo courtesy of P. J. Laughton).

modification is often associated with the plug flow configuration (see Chapter V, p. 5.7) of aeration tanks (length to width ratio of 5 to 50). A range of operating data from U. S. conventional activated sludge sewage treatment plants is presented in Table 4-2 (Environmental Protection Agency, 1974).

For conventional activated sludge systems the "Ten States Standards" suggest the aeration tank capacities and permissible loadings as presented in Table 4-3. These Standards call also for minimum air requirements of 1,500 cu. ft. per lb of BOD aeration tank load (93 cu. m. per kg BOD), and an average sludge recycle of 30 percent (15 percent minimum and 75 percent maximum).

Tapered aeration is a modification of the conventional activated sludge process, characterized by oxygen input tailored in accordance with the mixed liquor oxygen requirements (Haseltine, 1937; Kessler et al., 1936). Of course this modification is valid only for aeration tanks with a limited back mixing. The applied intensity of aeration is highest at the head of the tank, and diminishes with its length.

Efforts have been made to improve the conventional systems of the activated sludge process. Kraus (1945 & 1946) applied aerated digested sludge and aerated digester overflow to control bulking of activated sludge overloaded by seasonal canning wastewater. During the separate aeration of the

TABLE 4-2

Operating Data from Conventional Activated Sludge Plants
(after: Environmental Protection Agency, 1974)

Characteristics	Range of Values	
	From	To
Influent BOD, mg/L	92	254
Effluent BOD, mg/L	6	36
BOD Decrease, percent	77	96
Sludge Recycle, percent	16	52
Aeration Tank MLSS, mg/L	1,160	2,801
Aeration Detention Time*, hrs.	5.7	10
Air supplied, cu. ft/lb BOD removed	435	1,900
cu. m/kg BOD removed	27	118.4
Substrate loading, lb(g) BOD/lb(g)MLSS.day	0.15	0.45
Volume loading, lb BOD/day/1,000 cu. ft.	17	45
g BOD/day/cu.m	272	720

*excluding sludge recycle

digesting products, ammonia was converted into nitrates and solids were
converted into a good carrier for activated sludge bacteria. This way, the
aeration capacity of the original treatment system was improved as well as
the settling properties of the overloaded activated sludge. Other improvements of the conventional activated sludge were based on the addition of
mineral suspensions or some chemicals to the treatment system and these
are described in p. 4.10.

4.4 STEP-LOADING AND COMPLETELY-MIXED SYSTEMS

The basic idea behind a step-loading modification of the activated sludge
process (incorrectly called a step-areation system) is the application of
multiple wastewater feedpoints in the aeration tank, which makes the food to
micro-organism ratio in this system more uniform. This idea was developed
originally by Kessener in 1937 (de Jong, 1957), and applied in the treatment

TABLE 4-3
Conventional Tank Capacities and Permissible Loadings
(after: "Ten States Standards," 1973)

	Design Flow		
Requirements	to 0.5 mgd (1,893 cu.m/day)	0.5-1.5 mgd (1,893-5,678 cu.m/day)	1.5 mgd up (5,678 cu.m/day up)
Aeration retention, hrs.	7.5	7.5-6.0	6.0
Volume loading,			
lb BOD/1,000 cu. ft/day	30	30-40	40
g BOD/cu.m/day	480	480-640	640
Substrate loading,			
lb BOD/lb MLSS/ day	0.25-0.5	0.25-0.5	0.25-0.5
or			
g BOD/g MLSS/ day			

plants at Stockport, Vught, and Hilversum, Holland. Two years later, an identical approach was worked out by Gould (1939) in New York, and later developed by Torpey (1948). The step-loading plants are usually designed for aeration tank daily loadings of 50 to 60 lb BOD/1,000 cu. ft. (800 to 960 g/day.cu.m), with F/M ratios in the range of 0.2 to 0.4 lb (g) BOD/day/lb (g)MLSS. The maintained sludge age (solids retention times) values are similar to those of conventional systems. This design pattern provides for the addition of return sludge at the beginning of the aeration tank and the addition of wastewater in steps along the course of the mixed liquor flow (Fig. 4-2). In a typical four-pass aeration tank, with the load added at four various points, activated sludge is stored at higher concentrations toward beginning of the tank and is loaded and diluted successively by waste additions until, at the tank outlet, aeration tank effluent flows into the final settling tanks at a relatively lower mixed liquor suspended solids concentration (Fig. 4-3).

The range of operating data from step-loading plants is presented in

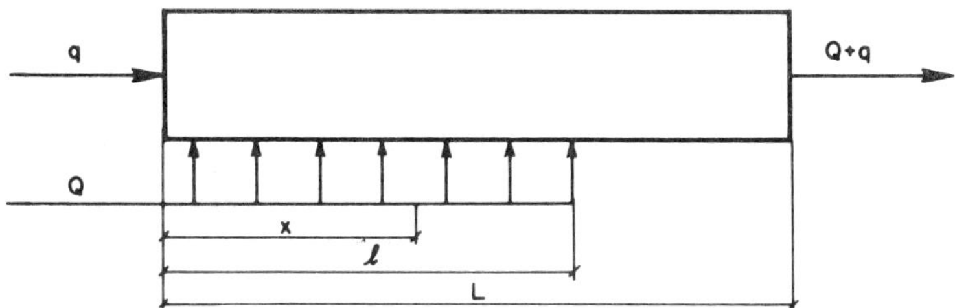

Figure 4-2. Step-Loading Modification (X/l = wastewater loading ratio for a given length of the aeration tank, L = total length of the aeration tank).

Table 4-4. According to the "Ten States Standards" (1973) this modification of the activated sludge process requires for aeration a minimum of 1,000 cu. ft. of air per lb. of BOD load (62 cu.m/kg) and a recycle of 50 percent (minimum 20 and maximum 75 percent). The suggested loadings for plants with a wastewater flow between 0.5 and 1.5 mgd (1,893 and 5,678 cu. m. per day) are from 30 to 50 lb BOD per day per 1,000 cu. ft. (480-800 g BOD per day per cu. m.). For larger plants the volume loadings of 50 lb BOD per day 1,000 cu. ft. (800 g BOD per day per cu. m.) are recommended. The respective substrate loadings are in the range of 0.2 - 0.5 lb (g) BOD per day per lb (g) MLSS, and the aeration retention (based on the design flow) is in the range of 5 to 7.5 hours.

A mathematical model of the step-loading (step-aeration) modification of the activated sludge process was developed by Weers and Andrews (1974). It was based on the assumption that the aeration tank was made up of several passes, each of which operated as a complete mixed reactor. The conventional activated sludge process was considered by these authors as a special example of the step-loading modification.

A further stage in the direction of step-loading is feeding wastewater to the aeration tanks along a substantial part of the length of its wall, or even as an overflow through this wall (Schmitz-Lenders, 1954). The completely-mixed systems are ultimate solutions for the idea of step-loading as a tool for achieving uniformity of sludge loading (food to micro-organisms ratios) in the activated sludge process (McKinney et al., 1958). This uniformity is a consequence of enough effective mixing in the aeration tanks to eliminate any concentration gradients of the process substrates, products, and of the microbial mass. Such an approach in the 1960's was met with considerable enthusiasm. A technological comparison of the performance of completely-mixed and plug-flow aeration tanks in Freeport, Ill. (Toerber, et al., 1974) showed that both systems yielded essentially identical removal efficiencies. However, in a response to a severe shock load, the completely-mixed system had an overall removal efficiency about 10 percent greater than the

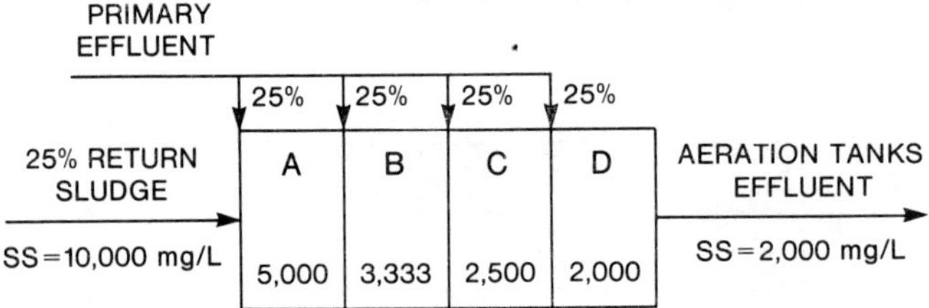

Figure 4-3. Example of step-loading operation (adapted from Torpey, 1948).

TABLE 4-4
Operating Data from Step-Aeration Plants
(after: Environmental Protection Agency, 1974)

Characteristics	Range of Values	
	From	To
Influent BOD, mg/L	74	140
Effluent BOD, mg/L	3	18
BOD Decrease, percent	84	94
Sludge Recycle, percent	16	92
Aeration Tank MLSS, mg/L	1,110	4,400
Aeration Detention Time*, hrs.	3.1	8.4
Air supplied, cu. ft/lb BOD removed	910	2,353
cu. m/kg BOD removed	57	147
Substrate loading, lb(g) BOD/lb(g)MLSS.day	0.1	0.54
Volume loading, lb BOD/day/1,000 cu. ft.	23	71
g BOD/day/cu.m	368	1,136

*excluding sludge recycle

plug-flow system. In the treatment of some industrial effluents (see
Chapter VIII) or mixed wastewater containing substances which do not bio-
degrade easily, completely-mixed systems may prove inferior in terms of
final effluent quality to limited-mixing conventional systems.

4.5 CONTACT STABILIZATION PROCESS

Although the concept of return sludge reaeration was practised as early as
the 1930's, the contact stabilization modification of the activated sludge
process was originally developed by Ullrich and Smith (1951, 1957) to solve
the problem of overloading at the sewage treatment plant in Austin, Texas.
The existing activated sludge system was designed for an average flow of
6.0 mgd (22,710 cu. m/day), but by 1950 the wastewater arriving at the
plant reached an average 10.663 mgd (40,400 cu. m/day). This innovation,
named "Biosorption" by its authors, was based on the diversion of the raw
sewage flow from the front end of the aeration tank to a contact zone near
the rear end of the tank (Fig. 4-4). It is also possible to design the reaeration zone as a separate reaeration tank (Fig. 4-5).

The results of the full-scale studies by Ullrich and Smith (1957) are
presented in Table 4-5. They described this new modification of the activated
sludge treatment as a process involving the adsorption and absorption of
pollutants in a short detention time (15 to 30 min.) contact phase followed
by settling. The sludge is then returned to a reaeration zone where the retained pollutants are processed by the micro-organisms prior to sludge
recycling to the contact zone. For years there has been controversy as to
whether the rapid initial uptake of BOD in the contact tank is due to physical
adsorption or is a biological phenomenon (see Chapter II, p. 2.1). The

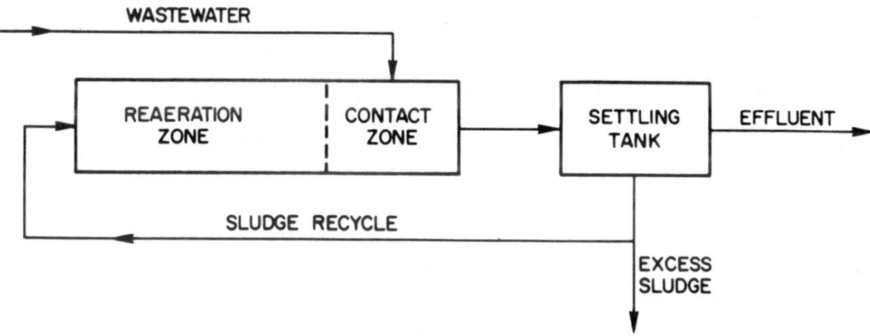

Figure 4-4. Contact stabilization process

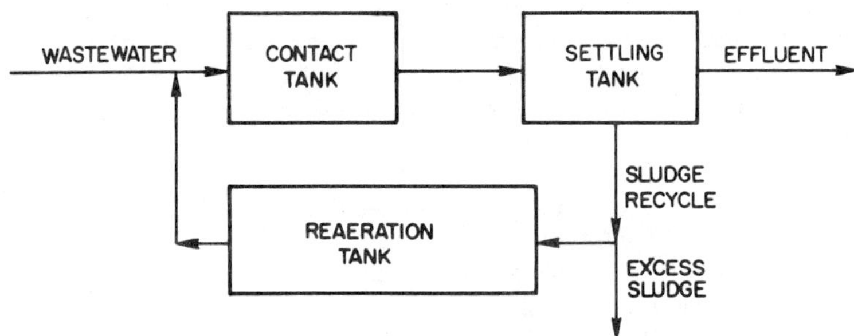

Figure 4-5. Separate sludge re-aeration in contact stabilization process.

substantial oxygen requirement associated with this stage of the treatment, supports decisively the biological theory. Jenkins and Orhon (1972) explained the mechanism of contact stabilization to be a rapid growth of microorganisms in the contact tank, accompanied by an increase in activated sludge viable fraction. While in the reaeration tank, cell death is predominant, accompanied by a decrease in viable fraction. This explanation complies with some previous observations by Siddiqi et al. (1967).

One advantage of the contact-stabilization modification is that the larger fraction of biological organisms is outside the main wastewater flow stream. Thus, if toxic conditions occur, the system can be returned to its normal operation more easily than with a conventional system.

For the design of the contact stabilization process, two techniques are usually used: one based on the food to micro-organism ratio, and the second on sludge age (Benefield and Randall, 1976). Jones (1970) proposed a mathematical model for contact stabilization based on a classical understanding of the process. The contact stabilization modification is broadly applied in package plant design for sewage treatment in small- and medium-sized communities (see Chapter IX, p. 9.6). A range of operating data from larger contact stabilization plants is presented in Table 4-6 (Environmental Protection Agency, 1974). This modification is capable of achieving only a partial nitrification of effluents. The "Ten States Standards" (1973) suggest, for contact stabilization, the same aeration tank volume loadings and substrate loadings as for the step-loading modification. The required contact zone aeration periods are presented in Table 4-7. According to these Standards, the contact zone should represent 30-35 percent of the total aeration capacity and the balance is for the reaeration. Moreover, the Standards state a minimum air requirement of 1,500 cu. ft. per lb BOD load (93 cu. m. per kg BOD) and an average sludge recycle of 100 percent (minimum 50 percent and maximum 150 percent). The long aeration time requirement in the contact tanks (Table 4-7) does not produce an optimal performance of this modification and perhaps is one of the reasons why many contact stabilization plants do not operate well.

TABLE 4-5
Biosorption Studies by Ullrich and Smith

Characteristic	May 1951	June 1951	January 1957	Averages
Biochemical oxygen demand (mg/L):				
Raw wastewater	229	198	358	262
Final effluent	15	9	21	15
Percent reduction	94	96	94	94
Suspended solids (mg/L):				
Raw wastewater	217	187	247	217
Final effluent	13	7	21	14
Percent removal	94	96	92	94
Return sludge	8,257	6,310	6,917	7,161
Reaeration phase	5,052	4,028	—	—
Contact phase	3,094	2,522	2,533	2,716
Return sludge (% of raw flow)	35.9	36.0	56.0	42.6
Sludge Volume Index (mL/g)	164	156	66	129
Detention time (raw flow basis):				
Contact tank (min)	31	31	38	33
Reaeration tank (hr)	2.7	2.75	5.2	3.5

Example   The operational data from the North Wastewater Treatment facility in Memphis, TN, can be presented as a current example of the contact stabilization process (Collins, 1982). This plant operated without primary clarifiers and with an average detention time in the contact tanks of 2.2 hrs. The respective MLSS concentration was 3,730 mg/L and F/M ratio equaled to 0.87. The stabilization tanks provided detention time of 6.5 hrs. for recycling sludge of MLSS concentration of 10,630 mg/L. The average SVI was 118 ml/g, and the sludge age was

equal to 4.5 d. The plant influent was composed approximately of 60%
industrial effluents and 40% domestic sewage. In 1980/1981 its average
BOD was 398 mg/L and TSS was 461 mg/L. Although this wastewater
contained relatively low concentrations of heavy metals, significant
concentrations of various pesticides, pesticide by-products, and solvents were commonly found in it. The treated effluent characterized
with an average BOD at 14.4 mg/L (over 96% removal), and TSS of 16.9
mg/L (removal of 86.3%).

4.6 HIGH-EFFICIENCY AND HIGH DISSOLVED OXYGEN SYSTEMS

High-efficiency activated sludge systems are characterized by relatively
high levels of mixed liquor suspended solids and high oxygenation capacity
of the aerators applied. These systems achieve relatively good treatment
results at high volumetric loadings. Even better treatment results can be
achieved in high dissolved oxygen systems by the application of commercial

TABLE 4-6
Operating Data from Contact Stabilization Plants
(after: Environmental Protection Agency, 1974.)

Characteristics	Range of Values	
	From	To
Influent BOD, mg/L	267	358
Effluent BOD, mg/L	17	30
BOD Decrease, percent	90	94
Contact tank MLSS, mg/L	1,377	4,000
Reaeration tank MLSS, mg/L	6,050	8,018
Contact tank detention time*, min.	45	100
Reaeration tank detention time**, min.	144	392
Substrate loading***, lb. (g) BOD/day/lb. (g) MLSS	0.32	0.49
Volume loading***, lb. BOD/day/1,000 cu. ft.	104	152
g BOD/day/cu. m	1,664	2,430

*based on influent flow excluding sludge recycle
**based on sludge recycle flow
***based on contact and reaeration volume

TABLE 4-7
Contact Zone Aeration Periods
(after: "Ten States Standards," 1973)

Design Flow		Aeration Period
mgd	cu. m/day	hrs.
to 0.5	to 1,893	3.0
0.5 - 1.5	1,893 - 5,678	3.0 to 2.0
1.5 up	5,678 up	1.5 to 2.0

oxygen, like the covered tank UNOX process, the "deep-shaft" method, and the fluidized bed process modification utilizing commercial oxygen.

HIGH-EFFICIENCY SYSTEMS

The possibilities of achieving relatively good treatment effects at high loadings of activated sludge systems, operating with conventional types of aerators, were first discussed by Wuhrmann (1953), and Pasveer (1954 and 1955). The first treatment plant designed according to this concept was described by Kehr and Schmidt-Bregas (1957). The high-efficiency systems apply volumetric loadings of aeration tanks in the range from 300 to 1550 lb BOD per 1,000 cu. ft. per day (5,000 - 25,000 g BOD.cu. m/day), sludge loadings to 2.7 lb (or g) BOD per lb (or g MLSS) per day, and MLSS contents in the range of 3,000 to 8,000 mg/L. The treatment results that can be achieved are in the range of 70 to 90 percent of BOD reduction, but the residual levels of suspended solids are often exceedingly high. Generally, activated sludge systems designed to operate at high mixed liquor suspended solids concentrations require comparatively smaller aeration tank volumes. However, this makes them more sensitive to input fluctuations, and the high mixed liquor suspended solids concentrations result in lower sludge settling velocities and therefore, larger secondary clarifiers are required for sludge separation and thickening.

The initial experience with the high-efficiency activated sludge system was described by Kehr and von der Emde (1960); the complementing data on two pilot-plants, operating at very short aeration times, are presented in Table 4-8.

OXYGEN SYSTEMS

Okun (1949) applied commercial oxygen to satisfy the oxygen uptake of activated sludge and showed that a high degree of wastewater treatment can be achieved at volumetric loadings surpassing 300 lb BOD per 1,000 cu. ft.

TABLE 4-8
High-Efficiency Treatment Pilot-Plant Experiments

Source of Information	Primary Effluent		MLSS mg/L	Aeration time, hr	BOD Reduction, %
	BOD, mg/L	SS, mg/L			
Benedek's experiments in Budapest,	207	170	7,700	0.50	89.2
Hungary	316	170	4,030	0.36	83.4
von der Emde's experiments in	141	n.d.	4,300	0.35	87.0
Kassel, Germany	141	n.d.	4,200	0.20	75.0

n.d. = not determined

per day (5,000 g BOD/cu. m./day). The major asset of oxygen systems appears to be their ability to obtain high oxygenation capacities in activated sludge reactors (aeration tanks) at relatively low turbulence levels. Moreover, to achieve the same treatment effects, these systems can be designed at higher F/M ratio levels than air systems, apparently because of higher availability of the mixed liquor microorganisms at a higher level of dissolved oxygen. Some oxygen systems require covered and staged tanks but others, for the price of some of the characteristic features, can be operated in open tanks (see Chapter V, p. 5.5). The latter, however, may not maintain higher levels of dissolved oxygen, and therefore, may not show any improved treatment abilities.

The results of the intensive development of the Union Carbide oxygen system, now under the name of UNOX, are available in a two volume monograph (McWhirter, 1978). The UNOX technology operates with an average 60% of oxygen in the commercial oxygen gas, which at the applied over-pressures of 2-4 in. (5-10 cm) W.C. may produce oxygen saturation values of 29 mg/L. Typical UNOX sewage treatment technology calls for dissolved oxygen levels in the aeration tanks in the range of 4-8 mg/L, mixed liquor suspended solids levels of 3,500 - 7,500 mg/L, and wastewater hydraulics detention times of 1-2 hours. The comparative values for air aerated systems (conventional modification of the process) are the dissolved oxygen levels of 1-2 mg/L, the mixed liquor suspended solids levels of 1,500 - 3,500 mg/L, and retention times of about 6 hours. It is claimed that the UNOX system offers space and energy efficiency and process dependability in handling peak loads. E.G. the Springfield, Missouri, sewage treatment plant, designed for an average flow of 30 mgd (113,550 cu. m/day) is capable of handling peak flows of up to 42 mgd (158,970 cu. m/day) without application of equalization basins. The results from a UNOX system pilot installation at Fall River, Massachusetts, for sewage treatment with about 30% industrial BOD load, showed that it was possible to meet the effluent requirements of 30 mg/L BOD and 30 mg/L suspended solids under average sludge loading conditions (F/M) of 0.60, with actual loadings fluctuating from 0.43 to 1.43. The excess activated sludge could be gravity thickened to 3 - 4.7% concentration at solids loading of 4.7 lg/day/sq. ft. (23 kg/day/ sq. m). Two examples of the application of the UNOX technology to the treatment of pulp mill effluents are presented in Chapter VIII (p. 8.5). The UNOX systems are particularly applicable where the available space for the treatment facilities is very limited, where the facilities have to be completely covered because of climatic reasons (Fig. 4-6) or because of the proximity of dwellings, and where a relatively strong wastewater is being treated.

DEEP SHAFT SYSTEM

<u>Deep-Shaft</u> system is based on a special design of the aeration reactor (see Chapter V, p. 5.7) leading to very high levels (average of 25 mg/L) of dissolved oxygen in mixed liquor. After treatment the mixed liquor, which

Figure 4-6.   UNOX System in Winnipeg, Manitoba, sewage treatment plant (photo courtesy of Union Carbide).

TABLE 4-9

Characteristics, Operating Parameters, and Treatment
Effects of Paris, Ontario Deep-Shaft Pilot Plant
(after Sandford and Chisholm, 1977)

Shaft Characteristics:

Depth:	508 ft. (155 m)
Diameter:	15-1/4 in. (39 cm)

Operating Parameters:

Residence time:	55 min.
MLSS:	6,000 mg/L
F/M:	1.0 d^{-1}
Shaft cycles:	10

Treatment Effects: (average readings)

BOD decrease:	from 141 mg/L to 24 mg/L
COD decrease:	from 514 mg/L to 89 mg/L
SS decrease:	from 217 mg/L to 29 mg/L

typically contains 10,000 mg/L of suspended solids, is over-saturated with
air; the mixed liquor suspended solids can be conveniently separated from
treated effluents by flotation. A pressure reactor for the activated sludge
process, studied some time ago in Chicago, can be considered a technological precursor for this system (Heil, 1970).

An example of an application of this system is the treatment plant for a
potato processing plant in Emlichheim, Germany (Bolton et al., 1976). This
plant was designed to treat 8,000 gal/hr (30 cu.m/hr) of wastewater with an
initial BOD of 2,000 mg/L. The shaft is 43 inches in diameter (1.09 m) and
328 feet deep (100 m). The resulting COD removal was over 90 percent.
The first pilot deep-shaft installation in North America was located in
Paris, Ontario, and was used to treat municipal sewage with a high proportion of textile mill effluents (Sandford and Chisholm, 1977). The characteristics, operating parameters and treatment effects at this plant are presented in Table 4-9. At present (1980), there are two Canadian plants operating according to deep-shaft technology: the 0.55 mgd (2,100 cu. m/d)
facility for Molson's brewery in Barrie, Ont., and the 0.6 mgd (2,300 m/d)
sewage treatment facility in Virden, Manitoba. A third plant in Portage la
Prairie, Manitoba is under construction. In the U.S. a deep-shaft
demonstration plant is operating in Ithaca, N.Y.

FLUIDIZED-BED SYSTEM

Fluidized-bed activated sludge process uses reactors (see Chapter V)
containing suspended sand particles coated with activated sludge-type
micro-organisms (Fig. 4-7).

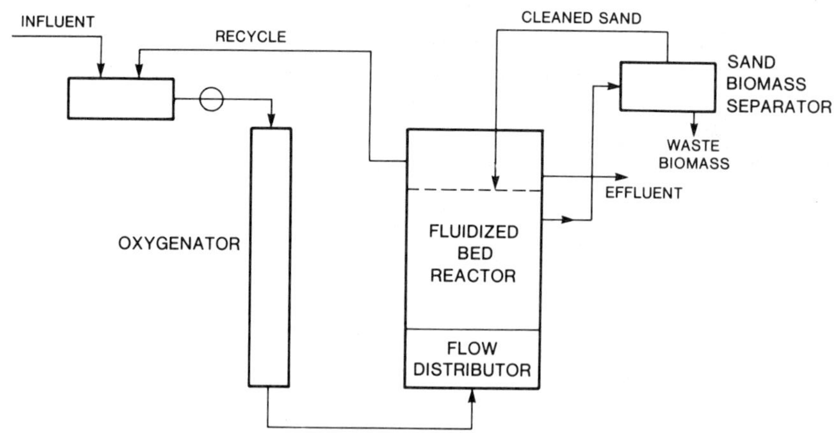

Figure 4-7. Biological fluidized bed treatment system.

Therefore, it can also be considered as a modification of biological film
processes. The fluidized-bed reactors are capable of retaining volatile
suspended solids concentrations in the range of 15 - 40 thousand mg/L.
This process allows throughput rates as low as a few minutes' retention
to achieve treatment effects comparable with a few hours' retention in the
conventional aeration tanks of the activated sludge process.

Originally, biological fluidized-bed reactors were used for denitrification under anoxic conditions (Jeris et al. 1974), and only later for aerobic
oxidation (Jeris et al. 1977). Subsequent developments of the system were
carried out by the Environmental Protection Agency, Dorr-Oliver (Sutton
et al., 1979), Canada Wastewater Technology Center, and in numerous other
places. Table 4-10 presents the performance results obtained at the Dorr-
Oliver pilot-plant in Orillia, Ont. (Dorr-Oliver, 1980). It is claimed that
the high biomass concentration in fluidized beds can reduce land area requirements for treatment by up to 80%; and markedly decrease capital costs
for the plant construction. Also the need for secondary clarification is eliminated or greatly reduced. However, many technological problems connected with the introduction of this system still exist. Experimental results
indicate that influent solids are carried through the process, the effectiveness
of oxygenation is often not satisfactory (and this calls for a substantial ratio
of recycle); separation of biomass from the carrier (mostly sand particles)
is far from being a well developed operation, etc.

4.7 PARTIAL TREATMENT SYSTEMS

The partial treatment or high rate activated sludge treatment systems, also
called "modified aeration" systems or "high-rate" systems, are characterized by a lower organics removal capacity than the process modification
described before (Setter, 1943 and 1944; Chase. 1944; Setter et al., 1945).
The basic technology for this modification depends on short-time (2-3)
hours of aeration wastewater, often without any pre-sedimentation of suspended solids, with only a recycle of 5-10% of sludge from secondary

TABLE 4-10
Fluidized-Bed Pilot-Plant Performance
Run #5, Orillia, Ont.
(after: Dorr-Oliver, 1980)

Characteristics	Values
Hydraulic Loading Rate, cu. m/sq. m/hr	20.3
Expanded Bed Height, m	3.66
Hydraulic Retention Time, min.	26
Temperature, °C	11
Recycle Ratio	3.1
Influent BOD, mg/L O_2	81
Effluent BOD, mg/L O_2	22
Influent Suspended Solids, mg/L	126
Effluent Suspended Solids, mg/L	37
Volatile Solids Concentration in Bed, mg/L	15,075
Sludge Produced, L/day	2,000

clarifiers. Under such conditions, mixed liquor susprnded solids are maintained at a level of 300-700 mg/L, and the sludge age in in range of 0.2-
0.5 days. The usual treatment effects for this process modification show a
wastewater BOD removal of the order of 55 to 80%. The results of full-
scale experiments on this modifiaction, performed by Torpey and Lang
(1958) at the treatment plant at Owls Head (New York City), indicated that
wastewater aeration for only one hour is the economic optimum. Previously
suggested extensions of the aeration time did not improve the treatment
effects. For "modified aeration" systems, the "Ten States Standards"
(1973) require much less air to be supplied (from 400 cu. ft. per lb BOD
load = 25 cu. m per kg), and an average sludge recycle of only 20 percent
(minimum 10 percent and maximum 50 percent). These Standards call also
for aeration retention of 2.5 hours and more, and suggest a volume loading
of 100 lb BOD per day per 1,000 cu. ft. (1,600 g BOD per day per cu. m)
with F/M equal to 1.0 or more. The operating data from the U.S. partial
treatment plants are presented in Table 4-11.

The so-called "activated aeration" was developed in New York City
(Chasick, 1954) for application in modified aeration systems. In this
The basic technology for this modification depends on a short-time (2-3
hours) aeration of wastewater, often without any pre-sedimentation of suspended solids, with only a recycle of 5-10% of sludge from secondary
clarifiers. Under such conditions, mixed liquor suspended solids are maintained at a level of 300-700 mg/L, and the sludge age is in the range of 0.2-
0.5 days. The usual treatment effects for this process modification show a

TABLE 4-11
Operating Data from Activated Sludge Partial Treatment Plants
(after: Environmental Protection Agency, 1974)

Characteristics	Range of Values	
	From	To
Influent BOD, mg/L	146	205
Effluent BOD, mg/L	36	66
BOD Decrease, percent	59	73
Aeration tank MLSS, mg/L	275	775
Sludge recycle, percent	6	13
Aeration Detention Time*, hrs.	1.5	2.5
Air supplied, cu. ft/lb BOD removed	639	1,171
cu.m/kg BOD removed	40	73
Substrate loading, lb(g) BOD/lb(g)MLSS.day	2.3	6.2
Volume loading, lb BOD/day/1,000 cu. ft.	93	138
g BOD/day/cu.m	1,493	2,215

*excluding sludge recycle

wastewater BOD removal of the order of 55 to 80%. The results of full-
scale experiments on this modification, performed by Torpey and Lang
(1958) at the treatment plant at Owls Head (New York City), indicated that
design, the excess activated sludge from another treatment system is used
as a source of micro-organisms for a system in which the sludge is not recycled (Fig. 4-8).

Further increases of F/M ratio, beyond the range used in modified
aeration systems, lead to formation of a dispersed bacterial growth. There-

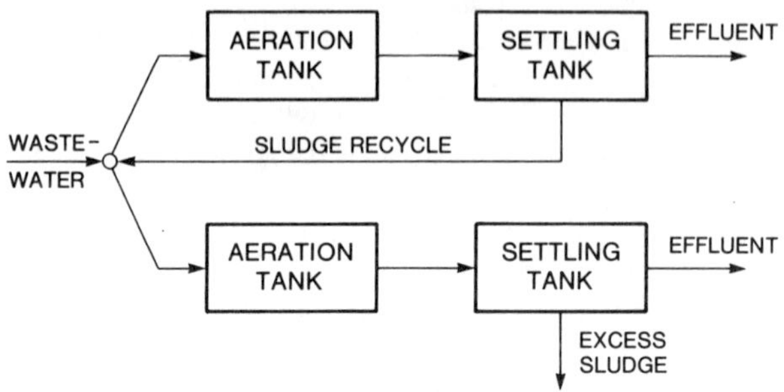

Figure 4-8. Activated aeration system.

fore, they are not applicable in the activated sludge process, as the
dispersed bacterial suspensions cannot be separated in secondary clarifiers.
However, aeration of wastewater with dispersed bacterial growth can be
considered as a pre-treatment stage for some concentrated industrial
effluents (Heukelekian, 1949; Nemerow, 1953 and 1954), with the condition
that some kind of separation of the dispersed bacterial growth is provided
by subsequent treatment stages.

4.8 TWO-STAGE SYSTEMS

An application of two-stage biological treatment usually shows better reliability, and a smaller production of excess sludge than a one-stage treatment
process (Imhoff, 1953; Lindner, 1957). Often the second stage biological
treatment is used to obtain improved nitrification of effluents, although, for
the majority of wastewater, it is possible to achieve such treatment in the
one-stage process. Moreover, an application of a two-stage process may
also help in a partial wastewater denitrification by recycling the second
stage effluent to the first stage treatment (see p. 4.9).

Two-stage biological treatment could be applied as a biological pretreatment with a biological filter, followed by an activated sludge system,
two-stage activated sludge system (Fig. 4-9), or a special combination of
treatment with a biological filter, coupled with an aeration tank like the so-
called "activated bio-filter". The coupled biological filter-activated sludge
systems are particularly adaptable for upgrading existing conventional
trickling filter plants, where more stringent treatment requirements have
been imposed, and where existing structures and equipment are in good condition (Stenquist et al., 1978). The trickling filter acts as a roughing filter
ahead of the activated sludge process, providing protection against various
upsets, and, especially, supporting the nitrification process which is more
sensitive than the carbonaceous oxidation.

Characteristic performance data for two-stage activated sludge treatment plants are presented in Table 4-12 (Environmental Protection Agency,
1974). These plants showed generally an excellent decrease in wastewater
BOD. Rimer and Woodward (1972) studied, in a pilot-plant scale, two-stage

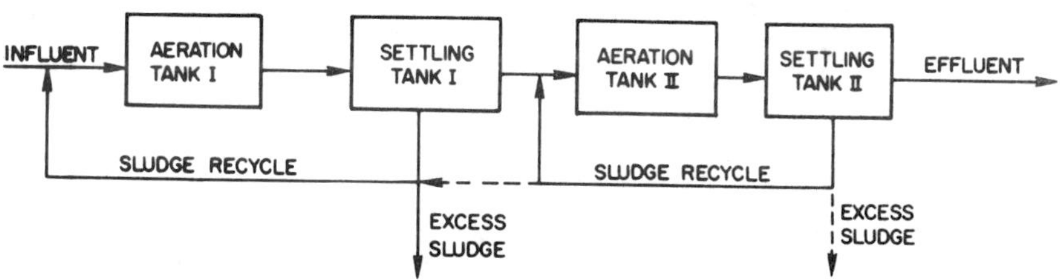

Figure 4-9. Two-stage activated sludge system.

TABLE 4-12
Operating Data from Two-Stage Activated Sludge Plants
(after Environmental Protection Agency, 1974)

Characteristics	Range of Values	
	From	To
1st-stage influent BOD, mg/L	104	271
1st-stage effluent BOD, mg/L	10	41
1st-stage BOD reduction, percent	81	96
1st-stage aeration tank MLSS, mg/L	1,920	3,760
1st-stage sludge recycle, percent	19	56
1st-stage aeration time*, hrs	0.7	2.9
1st-stage air supplied, cu ft/lb BOD reduced	820	1,180
cu m/kg BOD reduced	49	71
1st-stage substrate loading lb(g)BOD/day/lb(g)MLSS	0.92	1.56
1st-stage volume loading, lb BOD/day/1,000 cu. ft	128	375
g BOD/day/cu. m	2,048	6,000
2nd-stage effluent BOD, mg/L	4	21
2nd-stage BOD reduction, percent	16	79
2nd-stage aeration tank, MLSS, mg/L	800	1,600
2nd-stage sludge recycle, percent	11	29
2nd-stage aeration time*, hrs	0.7	2.9
2nd-stage air supplied**, cu. ft/lb BOD reduced	4,100	4,600
cu. m/kg BOD reduced	246	276
2nd-stage substrate loading, lb(g)BOD/day/lb(g)MLSS	0.20	0.59
2nd-stage volume loading, lb BOD/day/1,000 cu. ft	17.0	44.0
g BOD/day/cu. m	272	704
Overall BOD reduction, percent	93	96

*excluding sludge recycle

**including nitrification needs

activated sludge treatment of sewage containing about 30 percent of industrial
effluents. It was found that nitrification in the second stage was substantially
affected by temperature and wastewater pH. Specifically, no appreciable
nitrification was observed at temperatures lower than 50°F (10°C) and pH
higher than 8.3 - 8.5.

Zurn-Attisholz process is a modification of the two-stage activated
sludge system marketed in North America by Zurn Industries (Van Soest
and Das, 1973). It is claimed that, in this modification, the first stage
utilizes a mixed culture predominant in bacteria and lower fungi, and the
second stage - a culture predominant in protozoa. Under these circumstances, the dispersed growth carried over from the first stage settling is
retained in the second stage by predative protozoa which settle much better
than bacteria and produce a clear effluent. For operation of this process it
is suggested (Van Soest and Das, 1973) that, in the first-stage aeration
tanks, the mixed liquor suspended solids level be maintained at 8,000 -
15,000 mg/L, and the dissolved oxygen at 0.5 - 1.0 mg/L. At the design
BOD loading rate of up to 600 lb per day per 1,000 cu. ft. (9,600 g per day
per cu. m.), the BOD removal in this stage is 80 to 85 percent. The applicable sludge recycle ratios are 150 - 200 percent, and the first stage secondary clarifier operates at surface loading rates in the range of 500 - 1,500
gal. per day per sq. ft. (20 - 61 cu. m. per day per sq. m). The second
stage aeration requires loading rates up to 200 lb BOD per day per 1,000
cu. ft. (3,200 g per day per cu. m) and operates at mixed liquor suspended
solids of 500 - 1,000 mg/L and a dissolved oxygen level of 2 - 6 mg/L. It
is claimed that the total excess sludge production in the Zurn-Attisholz
process is only 0.4 lb (g) per lb (g) of BOD removed and the energy requirements are in the order of 0.40 - 0.45 HP-hr per lb BOD removed (0.66 -
0.74 kW hr per kg BOD removed).

Activated Bio-Filter system, developed by Neptune Microfloc Inc., can
be considered as a special design of a two-stage activated sludge system

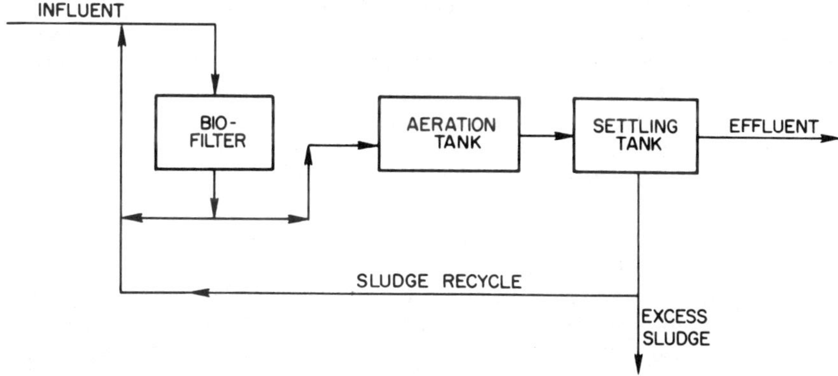

Figure 4-10. Activated bio-filter.

without an intermediate clarifier (a combined sludge system) and characterized by an internal recycle and an application of trickling filtration principle
for mixed liquor solids retention and aeration in the first-stage treatment
(Fig. 4-10). In this treatment, wastewater is combined with recycling activated sludge and the mixed liquor is distributed by spray nozzles over the
media of a biological filter (called bio-cell) consisting of evenly spaced,
horizontal redwood slats. A part of the filter underflow is recycled and the
balance is directed to an aeration tank followed by a secondary clarifier.
The idea of mixed liquor aeration in a roughing filter has been explored before in the "Spiro-vortex" system (Nelson, 1958), but the above design appears to be more practical. General design parameters for treating domestic sewage in such systems, with or without nitrification, were presented by
Owen and Slechta (1975).

4.9 NITRIFICATION-DENITRIFICATION SYSTEMS

Nitrification in the activated sludge process can be accomplished in a
single-stage combined (single) sludge treatment (Fig. 4-11), and a two-
stage separate sludge treatment (Fig. 4-8). It had been assumed originally
that the sludge separation was beneficial for the nitrification performance.
However, a combined sludge system proved to be a viable alternative for
wastewaters which did not contain nitrification inhibitors (Sutton et al.,
1975). Another assumption that combined sludge systems produce a better
flocculating sludge, resulting in a lower residual suspended solids concentration, was not confirmed by the above authors who found, in pilot-plant
experiments, that the mean clarifier effluent suspended solids concentration
was 24 mg/L in both systems. The residual filterable nitrogen in these
effluents was about 1 mg/L, with another 1 mg/L of non-filterable nitrogen.
The nitrification process is sensitive to temperature decreases. However,
this sensitivity decreases at high sludge retention times, which can be explained by a decrease in activation energy in the Arrhenius equation with the
increase of solids retention time. A slower response to temperature decrease was observed in the separate sludge systems than in combined sludge
systems (Sutton et al., 1975). The total excess of sludge is higher in the
separate systems than in the combined systems (Sutton and Jank, 1975).

For sewage nitrification in a separate nitrification stage it is recommended, that the nitrification tanks operate at a mixed liquor suspended
solids concentration higher than 2,000 mg/L. However, sometimes it may
be difficult to achieve this level. For this purpose sufficient organic carbon
concentration in the influent to the second stage unit has to be maintained,
to allow for the control of the sludge retention time (sludge age), i.e. the
sludge yield in the treatment must exceed the loss of sludge as the final
effluent suspended solids. The optimal rate of oxidation of 0.15 gNH_3/g

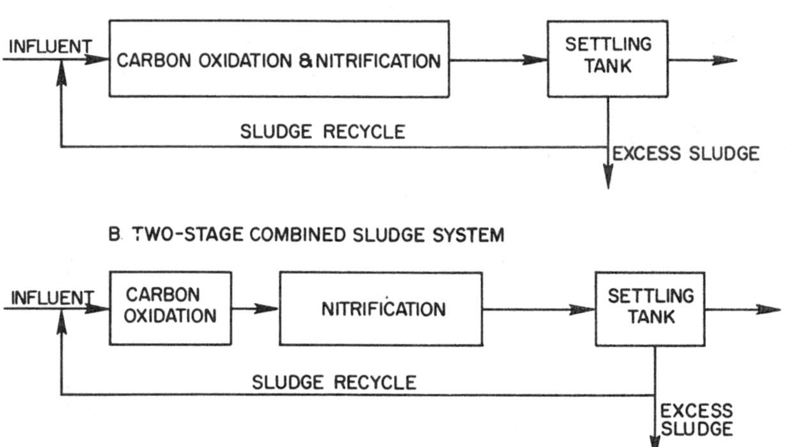

Figure 4-11.   Combined Sludge Single-Stage and Two-Stage Nitrification.

MLSS per day may be expected at 20°C. Separate nitrification stages are often characterized by very long sludge retention times (sludge age). For sewage nitrification, these values range from 15 to 25 days (Parker et al., 1975).

Denitrification can be obtained, using the same sludge which provided carbonaceous biological oxidation in stirred reactors (sequencing reactors) (Fig. 4-12) or using a separate denitrifying sludge. In the first approach, an internal carbon source is required for the process and, in the second one anexternal source of organic carbon may be utilized to a degree depending on the C/N ratio. The classical approach, called the three-sludge system (Fig. 4-13), implies first a removal of organic matter and then a separate nitrification, followed by denitrification under anoxic conditions, with controlled supplementation of a carbon source (hydrogen donors) for the denitrifiers.

The single sludge (combined sludge) denitrification systems (Fig. 4-14 and 4-15) were investigated by Ludzack and Ettinger (1962), Wuhrmann (1964), Christensen et al. (1975), etc. However, this process does not produce a consistently high nitrogen removal and requires long retention time. On the other hand, separate sludge denitrification with an external carbon source is very effective and is characterized by high reaction rates (Mulbarger, 1971; Horstkotte et al., 1974; Murphy and Sutton, 1974, etc.).

Separate denitrification reactors should be followed by aerobic stabilization chambers to reduce leakage of residual external source of organic carbon (methanol) from the process, and to prevent odour formation in secondary clarifiers.

Figure 4-12. Stirred Reactor for Denitrification (photo Author).

        Several practical designs of sequencing reactors have been offered for
elimination of nitrogen from wastewater through nitrification and denitrification. The removal of 75 to 84 percent of nitrogen from primary effluents
was reported in a single stage activated sludge process by an application of
a two-pass reactor (Bishop et al., 1974). At a loading of approximately
0.1 g BOD/g MLSS per day, air aeration was applied on a 30-minute cycle,
alternatively to the first and then to the second reactor pass. Mechanical
mixers suspended the mixed liquor solids when the air was not applied.
Effective elimination of nitrogen from wastewater was also obtained in a
system composed of four completely mixed basins in series followed by a
settling tank (Barnard, 1975). The first and the third basins were stirred
without addition of oxygen while the second and fourth basins were aerated.
Mixed liquor was partially recycled from the second basin to the first and
sludge from the settling tank was recycled to the first basin. Total nitrogen
removals of 94 percent were achieved this way. Another denitrification

possibility is with the operation of a nitrifying activated sludge process at a high level of mixed liquor suspended solids and/or high concentration of treated wastewater. Murray et al. (1975) observed, in an oxidation ditch treatment of animal farm waste, a substantial elimination of wastewater nitrogen (90 percent) by simultaneous nitrification-denitrification.

The basic problem with utilization of the sequencing reactors for nitrogen elimination appears to be sludge bulking. A more detailed presentation of problems associated with single sludge nitrogen elimination is covered by Sutton et al. (1979a) and Heidman (1979).

Figure 4-13. Three-sludges treatment system.

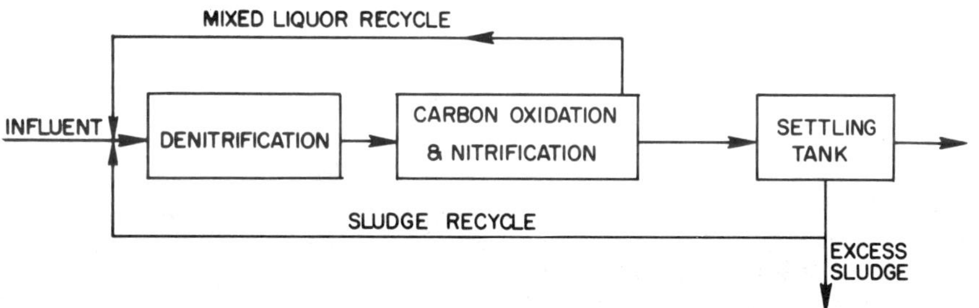

Figure 4-14. Ludzack-Ettinger Nitrogen Elimination System.

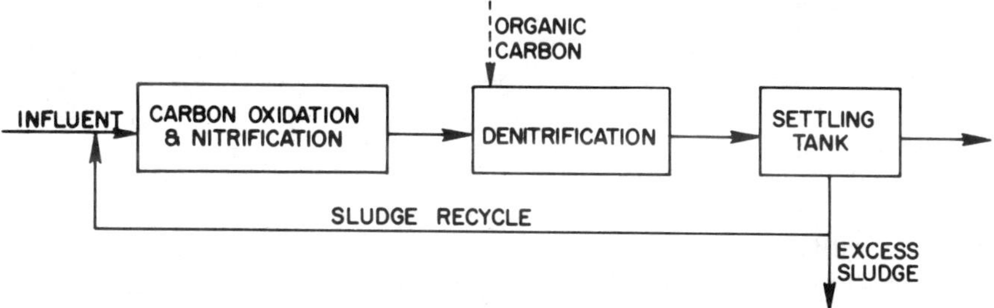

Figure 4-15. Two-stage, single-sludge (combined sludge) nitrification-denitrification system.

4.10 BIOLOGICAL-CHEMICAL SYSTEMS

Research into the integration of activated sludge process treatment with some kind of chemical precipitation was originally aimed at improved treatment in a conventional activated sludge process and — later on — was directed to also achieve effective precipitation of phosphates from wastewater. The latter purpose has gained some popularity because of the increasing requirements to prevent eutrophication of receiving water bodies.

The "Niersverfahren" in Germany and the Guggenheim process in the U.S.A. (Jung, 1939/1940, and Phelps and Bevan, 1942), were developed as early as the late 1930's. Both these processes were based generally on an addition of iron salts to wastewater prior to its activated sludge treatment; they proved quite successful in the treatment of municipal wastewaters and some industrial wastewaters. Dawson and Jenkins (1950) found that the ferrous ion additions to activated sludge caused a suppression of oxygen uptake in concentrations of 1,000 mg/L but had very little effect at levels of 100 mg/L. Intermediate concentrations of iron produced an increase in oxygen uptake, thus indicating that iron might be useful in increasing biological activity. Also, Carter and McKinney (1973) have shown that a higher rate of metabolism in the activated sludge process could be achieved when a proper balance of iron and other nutrients is made available to the microorganisms.

In 1973, Anderson and Hammer observed that alum addition to the activated sludge process adversely affected higher forms of micro-organisms, but did not affect BOD removal. In the microbiological studies of Unz and Davis (1975) it was also found that the activated sludge containing aluminum hydroxide showed significantly higher numbers of lipolitic, gelatinolytic and thiosulfate-oxidizing micro-organisms, and possibly fewer nitrite-oxidizing micro-organisms, than the control activated sludge. Amoeboid and ciliated protozoa, but not flagellates, were found less frequently in the alum-dosed aeration tank than in the control mixed liquor. Similarly, Lin and Carlson (1975) found that higher dosages of alum to activated sludge caused damage to the higher forms of micro-organisms and were capable of removing the filamentous growth from activated sludge systems.

Humenick and Kaufman (1970) developed a system involving two stages of treatment: an activated sludge process stage, followed by a precipitation-flocculation stage, with organic chemical solids recycled to the first stage to achieve continuous control over the separability of the biologically active solids. The integrated biological-chemical process offers economics in the removal of phosphorus and BOD from municipal wastewater. Lime is the chemical to be preferred in such installations because of its lower cost and the potential for its reuse. In domestic wastewater treatment, removals of BOD in excess of 90 percent and COD in excess of 80 percent were achieved at the volumetric loadings of the biological part of the system in the range of 400 to 500 lb BOD per day per 1,000 cu. ft. (6,400 to 8,000 g BOD per day

per cu. m). The system operated at a mixed liquor concentration exceeding
9,000 mg/L and the sludge volume index was less than 100 mg/L. Phosphorus removal was at least 95 percent.

Similar studies were also carried out by Stukenberg (1971) who investigated a system composed of high-rate activated sludge treatment, followed
by lime-precipitation and wastewater recarbonation. A 2.5 hr aeration,
followed by lime coagulation, produced effluent quality of about 4 mg/L BOD.
The application of such an approach for the upgrading of existing facilities
was suggested.

Minton and Carlson (1972) discussed thoroughly several aspects of different combined chemical-activated sludge treatments, including alkalinity
equilibria and pH limitations.

4.11 POWDERED ACTIVATED CARBON-ACTIVATED SLUDGE SYSTEMS

In 1972, Hutton reported that additions of powdered activated carbon (PAC)
to the activated sludge system used for an industrial wastewater treatment
helped to improve the treatment effects over the possibilities of activated
sludge process alone. Powdered activated carbon dosages of about 50 ppm
and over were introduced to the aeration tank, and bacteria grew on the
carbon to produce a coherent mass. They also provided a coagulating agent
for the activated carbon, causing it to settle quite well. Kalinske (1972)
hypothesized that biological oxidation in such systems was enhanced perhaps
through provision of highly localized organic substance concentrations and
growth sites for biomass.

The experiments of Scaramelli and DiGiano (1973), carried out at dosages of 100 and 200 mg/L of powdered activated carbon (Darco DXL-0-
5077), confirm the unusual effectiveness of this process in the removal of
residual organic substances from wastewater. It was observed that it was
not necessary to continuously add powdered activated carbon to the treatment facilities, and that such treatment caused a decrease in the sludge
volume index.

Additions of powdered activated carbon to activated sludge treatment
of concentrated dye wastewater proved very effective in removing colour
from the wastewater (LaRocca et al., 1975). This system utilized about 40
to 60 percent less carbon than a comparable once-through carbon adsorption
system, suggesting the possibility that carbon was partially regenerated in
the biological system. Biological degradation of some dye molecules seemed
to be more possible in this system than in the basic activated sludge process,
due to the long carbon solids retention time which permitted the adsorbed
molecules to remain under biological treatment for an extended period of
time. In another study (DeWalle et al., 1975) it was found that additions of
powdered activated carbon to activated sludge systems caused an improvement in the interface subsidence rates of the sludges having lower sludge
ages, but did not affect those having higher sludge ages. The removal of

residual organic matter in these systems generally increased with increasing dosages of powdered activated carbon.

Full-scale experiments on powdered activated carbon additions
(Hydrodacro H) at Norfolk, Nebraska (Spady and Adams, 1974) showed a
decrease in the suspended solids level in the final effluent, a decrease in
the sludge volume index and its variability, an improvement in wastewater
nitrification, and a substantial improvement in the filter yield in the sludge
vacuum filtration at a lower dose of cationic polymer. However, some
other full-scale studies of the application of powdered activated carbon to
conventional activated sludge treatment were not a technological success
(Averill et al., 1975). In the Bolton, Ontario, Sewage Treatment Plant,
doses of 25 and 50 mg/L of the powdered activated carbon Aqua Nuchar A
were added, with a dose of 150 mg/L alum, to the mixed liquor prior to
secondary clarification. Such treatment resulted in an improved removal
of soluble organic material measured as filtered BOD and COD, but caused
an increase in residual suspended solids in the final effluent. However, the
mixed liquor containing powdered activated carbon settled more rapidly
and produced a lower sludge volume index.

Similar treatment results to those with the powdered activated carbon-
activated sludge process have also been reported for an activated sludge
process assisted by the addition of spent oil-cracking catalyst (Schwarz and
McCoy, 1976). This spent catalyst contained alumina and alumina-silicates
which are known to have the ability to adsorb organic substances and micro-
organisms. Addition of the catalyst increased the activated sludge settling
rate, decreased the settled sludge volume, decreased the effluent residual
organic carbon and increased effluent clarity. It also helped to decrease
the phosphate contents in the treated wastewater.

4.12 BIOLOGICAL PHOSPHORUS REMOVAL

Modifications of the activated sludge process for the effective removal of
phosphorus during treatment of wastewater are based on the mechanism of
phosphorus uptake and release by activated sludge, described in more detail in Chapter II (p. 2.9). This mechanism is not yet fully understood. In
a number of activated sludge treatment plants, a substantial removal of
phosphorus has been reported, surpassing the treatment requirements.
Milbury et al., (1971) performed a thorough full-scale study at the
Baltimore, Maryland, conventional activated sludge plant to maximize the
phosphorus uptake from wastewater. An elimination of more than 80 percent of phosphorus from wastewater to the residual contents of 0.2 to 0.3
mg/L of P as soluble phosphates and 1.0 to 1.5 mg/L of P as total P,
proved possible by operating activated sludge aeration in a plug-flow configuration of at least 1,200 to 1,300 mg/L mixed liquor suspended solids
level, and maintaining the aeration so that dissolved oxygen was not

limiting. These effects depended on proper handling of phosphorus-rich activated sludge to prevent phosphorus release and recycle to the treatment system. This was achieved by a continuous wasting of excess sludge and avoiding anaerobic conditions in secondary clarifiers. The Union Carbide pilot-plant experiemtns in Findlay, Ohio, demonstrated a phosphorus removal from 8 mg/L in the plant influent to effluent total phosphorus concentrations of 0.7 mg/L, and soluble phosphorus concentrations of 0.1 mg/L.

Controlled phosphorus release from activated sludge and its subsequent disposal can be accomplished by several technological methods. Levin et al., (1975) described full-scale operation of a system in which settled activated sludge became anaerobic in an additional tank and produced a phosphorus-rich supernatant. Lime was used to precipitate the phosphorus from the supernatant. Economical aspects of such phosphorus removal at Reno-Sparks, Nevada, were analyzed by Peirano (1977), and a detailed analysis of the results achieved at this installation was presented by Drnevich (1979). This technology is marketed under the registered name of the "Phostrip" process.

4.13 THERMOPHILIC ACTIVATED SLUDGE

The thermophilic activated sludge process was the subject of studies by Hausmann and Malz (1959), Zuelke (1969), Shindala and Parker (1970), Surucu et al., (1976) and others. Although it is theoretically possible to operate such a process, the costs and technical problems associated with its operation for sewage treatment under normal conditions, outweigh the benefits. Zuelke (1969) reported that the most favourable ranges for operating a temperature adapted thermophilic activated sludge system were from 116 to 128°F (47 to 53°C) and from 137 to 149°F (58 to 65°C). For an unadapted mesophilic system, the upper limit for temperature was around 104°F (40°C). The thermophilic systems have to be closely controlled within the chosen temperature range because the temperature fluctuation leads to a substantial decrease in the metabolic capacity of micro-organisms. Shindala and Parker (1970) carried out an evaluation of an activated sludge process at 131°F (55°C) under varying organic loadings and retention times. Without sludge return, the overall BOD reduction was 94 percent. At a sludge recycle to influent ratio of 1:1, the average BOD reduction was 85 percent. Variation of MLSS from 200 to 650 mg/L had no significant effect on the removal efficiency. The SVI was 40 to 60 mg/L with no sludge return and 140 to 150 mg/L with sludge return. Surucu et al. (1976) studied aerobic thermophilic treatment of high-strength wastewater. It was shown that this process could be used to treat such wastewaters more efficiently, and at the same time produce less sludge than normally would be expected from the mesophilic system. It was also shown that a thermophilic system could be maintained in a temperature range of 50° to 60°C through the heat generated by microbial thermogenesis while treating high-strength wastewaters.

Application of elevated temperatures (86° to 95°F or 30° to 35°C) to activated sludge treatment of coal carbonization effluents has been suggested by several authors and has been implemented in a few treatment plants (see Chapter VIII, p. 8.4).

Duke et al. (1980) reported on problems arising from a full-scale treatment of Kraft pulp mill effluents and organic chemical plant effluents in an oxygen activated sludge system at temperatures in excess of 115°F (46°C). The study showed that the rate of BOD decrease was depressed at temperatures greater than 105°F (40.6°C) and the rate of solids separation in secondary clarifiers was adversely affected at temperatures greater than 100°F (37.8°C). Similar problems are also described in Chapter III (p. 3.9) and Chapter VIII (p. 8.5).

REFERENCES

Anderson, D. T., and Hammer, M. J.: "Effects of Alum Addition on Activated Sludge Biota," Water and Sewage Wks., 120, 1, 63 (1973).

Averill, D. W., et al.: "The Use of Activated Carbon in Conventional Activated Sludge Processes," paper presented at the Seminar on High Quality Effluents, Toronto, 1975.

Baars, J. K.: "The Use of Oxidation Ditches for Treatment of Sewage from Small Communities," Bull. World Health Org., 26, 465 (1962).

Barnard, J. L.: "Biological Nutrient Removal Without the Addition of Chemicals," Water Research (Brit.), 9, 485 (1975).

Benefield, L. D., and Randall, C. W.: "Design Procedure for a Contact Stabilization Activated Sludge Process," Jour. Water Poll. Control Fed., 48, 147 (1976).

Bishop, D. F., et al.: "Single Stage Nitrification," paper presented at the 47th Annual Conference of Water Pollution Control Federation, Denver, Col., October, 1974.

Bolton, D. H., et al.: "The Application of the ICI Deep Shaft Process to Industrial Effluents," paper presented at the 31st Purdue Industrial Waste Conference, Purdue University, Lafayette, Ind., 1976.

Burkhead, C. E., and Wood, D. J.: "Analog Simulation of Activated Sludge Systems," Proc. Am. Soc. Civ. Engrs., Jour. Sanit. Eng. Div., 95, SA3, 593 (1969).

Carter, J. L., and McKinney, R. W.: "Effects of Iron on Activated Sludge Treatment," Proc. Am. Soc. Civ. Eng., Jour. Envir. Eng. Div., 99, EE2, 135 (1973).

Chase, E. S.: "High-Rate Activated Sludge Treatment of Sewage," Sew. Wks. Jour., 16, 878 (1944).

Chasick, A. H.: "Activated Aeration at the Wards Island Sewage Treatment Plant," Sewage Ind. Wastes, 26, 1059 (1954).

Christensen, M. H., et al.: "Combined Sludge Denitrification of Sewage Utilizing Internal Carbon Sources," paper presented at the IAWPR Conference on Nitrogen as a Water Pollutant, Copenhagen, Denmark, 1975.

Collins, J. R., Jr.: "North Wastewater Treatment Facility, Memphis, Tennessee," Jour. Water Poll. Control Fed., 54, 229 (1982).

Dawson, P. S. S., and Jenkins, S. H.: "The Oxygen Requirements of Activated Sludge Determined by Manometric Methods. II. Chemical Factors Affecting Oxygen Uptake," Sewage and Ind. Wastes, 22, 490 (1950).

DeWalle, F. B., et al.: "Organic Matter Removal by Powdered Activated Carbon Added to Activated Sludge Units," paper presented at the 48th Annual Conference of Water Pollution Control Federation, Miami Beach, Fla., October, 1975.

Dorr-Oliver Inc.: information released in 1980.

Drnevich, R. F.: "Biological-Chemical Process for Removing Phosphorus at Reno/Sparks, NV," EPA-600/2-79-007, 1979.

Duke, M. L., et al.: "Evaluation of Problems in Operation of the High Temperature Pure Oxygen Activated Sludge Process," paper presented at the Industrial Waste Conference, Purdue University, 1980.

Environmental Protection Agency: "Process Design Manual for Upgrading Existing Wastewater Treatment Plants," Washington, D.C., 1974.

Gould, R. H.: "Talmans Island Opens for Worlds Fair," Munic. Sanit., 10, 185 (1939).

Great Lakes - Upper Mississippi River Board of State Sanitary Engineers: "Recommended Standards for Sewage Works," Albany, N.Y., 1973.

Haseltine, R. T.: "Operating Control Tests for the Activated Sludge Process," Waterworks and Sewerage, 84, 121 (1937).

Heidman, J. A.: "Sequential Nitrification-Denitrification in a Plug-Flow Activated Sludge System," EPA-600/2-79-257, 1979.

Heil, R. W.: personal information (1970).

Heukelekian, H.: "Aeration of Soluble Organic Waste with Nonflocculent Growth," Ind. Eng. Chem., 41, 1412 (1949).

Horstkotte, G. A., et al.: "Full-Scale Testing of a Water Reclamation System," Jour. Water Poll. Control Fed., 46, 181 (1974).

Humenick, M. J., and Kaufman, W. J.: "An Integrated Biological-Chemical Process for Municipal Wastewater Treatment," paper presented at the 5th International Water Pollution Research Conference, San Francisco, California, 1970.

Husmann, W., and Malz, F.: "Untersuchungen zur Biologischen Abwasserreinigung auf Aerober Thermophilic Grundlage," Gas-Wasser Fach (Ger.), 89, 100, 8, 189 (1959).

Hutton, D.: "Improved Biological Waste Water Treatment," DuPont Innovation, 3, No. 6 (1972).

Imhoff, K.: "Das Zweistufige Belebungsverfahren fur Abwasser," Gas und Wasserfach, 96, 43 (1953).

Jenkins, D., and Orhon, D.: "The Mechanism and Design of the Contact Stabilization Activated Sludge Process," paper presented at the 6th Int. Water Poll. Research Conf., Jerusalem, Israel, 1972.

Jeris, J. S. et al.: "High-Rate Biological Denitrification using a Granular Fluidized Bed," Jour. Water Poll. Control Fed., 46, 2118 (1974).

Jeris, J. S., et al.: "Biological Fluidized-Bed Treatment for BOD and Nitrogen Removal," Jour. Water Poll. Control Fed., 49, 816 (1977).

Jones, P. H.: "A Mathematical Model for Contact Stabilization Modification of the Activated Sludge Process," paper presented at the 5th Int. Water Poll. Research Conf., San Francisco, Calif., 1970.

de Jong, J.: "Het Bedrijf van de Beluchtingstanks van de Biologische Zuiveringsinstallatie voor het Leijgebied in de Gemeente Tilburg," Publieke Werken (Holand), No. 5, 47 (1957).

Jung, H.: "Praktische Erfahrungen mit dem Niersverfahren bei der Reiningung gewerblicher Abwasser," Vom Wasser 14, 216 (1939/1940).

Kalinske, A. A.: "Enhancement of Biological Oxidation of Organic Wastes Using Activated Carbon in Microbial Suspensions," Water and Sewage Wks., 116, 6, 62 (1972).

Kehr, D., and Schmidt-Bregas, F.: "Die neue Klaeranlage der Stadt Detmold," Bauamt und Gemeindebau (Ger.), 30, 240 (1957).

Kehr, D., and von der Emde, W.: "Experiments on the High-Rate Activated Sludge Process," Jour. Water Poll. Control Fed., 32, 1066 (1960).

Kessler, L. H., et al.: "Tapered Aeration of Activated Sludge," Munic. Sanit., I, 268 (1936).

Kraus, L. S.: "The Use of Digested Sludge and Digester Overflow to Control Bulking Activated Sludge," Sewage and Ind. Wastes, 17, 1177 (1945).

Kraus, L. S.: "Digested Sludge — An Aid to the Activated Sludge Process,"
Sewage and Ind. Wastes, 18, 1099 (1946).

LaRocca, S. A., et al.: "Treatment of Concentrated Dye Wastes by
Powdered Activated Carbon in Activated Sludge," paper presented at the
48th Annual Conference of Water Pollution Control Federation, Miami
Beach, Fla., October, 1975.

Levin, G. V., et al.: "Operation of Full-Scale Biological Phosphorus Removal Plant," Jour. Water Poll. Control Fed., 47, 569 (1975).

Lin, S. S., and Carlson, D. A.: "Phosphorus Removal by the Addition of
Aluminum (III) to the Activated Sludge Process," Jour. Water Poll. Control
Fed., 47, 1978 (1975).

Lindner, W.: "Das Zweistufige Belebungsverfahren in der Abwasserreinigung", Thomas Verl., Kempen-Niederrheim 1957.

Ludzack, F. J., and Ettinger, M. B.: "Controlling Operation to Minimize
Activated Sludge Effluent Nitrogen," Jour. Water Poll. Control Fed., 34,
920 (1962).

McKinney, R. E., et al.: "Design and Operation of a Complete Mixing
Activated Sludge System," Sewage Ind. Wastes, 30, 287 (1958).

McKinney, R. E.: "Mathematics of Complete Mixing Activated Sludge,"
Proc. Amer. Soc. Civ. Engrs., Jour. Sanit. Eng. Div., 88, SA3, 87
(1962).

McWhirten, J. R. (editor): "The Use of High-Purity Oxygen in the Activated
Sludge Process," Volume I and II, CRC Press, Inc., West Palm Beach,
Florida, 1978.

Milbury, W. F., et al.: "Operation of Conventional Activated Sludge for
Maximum Phosphorus Removal," Jour. Water Poll. Control Fed., 43,
1980 (1971).

Minton, G. R., and Carlson, D. A.: "Combined Biological-Chemical
Phosphorus Removal," Jour. Water Poll. Control Fed., 44, 1736 (1972).

Mulbarger, M. C.: "Nitrification and Denitrification in Activated Sludge
Systems," Jour. Water Poll. Control Fed., 43, 2059 (1971).

Murphy, K. L., and Sutton, P. M.: "Pilot Scale Studies on Biological
Denitrification," paper presented at the 7th International Conference on
Water Pollution Research, Paris, France, 1974.

Murray, I., et al.: "Inter-relationship Between Nitrogen Balance, pH and
Dissolved Oxygen in an Oxidation Ditch Treating Farm Animal Waste,"
Water Research (Brit.), 9, 25 (1975).

Nelson, F. G.: "Spiral Contact Aeration in Biological Treatment," Sewage
and Ind. Wastes, 30, 907 (1958).

Nemerow, W. L.: "Oxidation of Cotton Kier Wastes," Sewage Ind. Wastes, 25, 1060 (1953) and 26, 1231 (1954).

Owen, W. F., and Slechta, A. F.: "Organic Removal or Nitrification with a Combined Fixed-Suspended Growth Biological Treatment System," paper presented at the 48th Annual Conference of the Water Pollution Control Federation, Miami Beach, Fla., 1975.

Okun, D. A.: "System of Bio-Precipitation of Organic Matter from Sewage," Sewage Wks. Jour., 21, 763 (1949).

Parker, D. S. et al.: "Process Design Manual for Nitrogen Control," Technology Transfer, Environmental Protection Agency, October 1975.

Pasveer, A.: "Research on Activated Sludge," Sewage Ind. Wastes, 26, 149 (1954) and 27, 783 (1955).

Pasveer, A.: "Eenvoudige Afvalwaterzuivering," De Ingenieur (Holland), 67, No. 17, G1-6 (1957).

Pasveer, A.: "Abwasserreinigung im Oxydationsgraben," Bauamt und Gemeindebau (Ger.), 31, 78 (1958).

Peirano, L. E.: "Low-Cost Phosphorus Removal at Reno-Sparks, Nevada," Jour. Water Poll. Control Fed., 49, 568 (1977).

Phelps, E. B., and Bevan, I. C.: "Laboratory Study of the Guggenheim Biochemical Process," Sewage Works Jour., 14, 104 (1942).

Rimer, A. E., and Woodward, R. L.: "Two-Stage Activated Sludge Pilot-Plant Operations at Fitchburg, Mass.", Jour. Water Poll. Control Fed., 44, 1, 101 (1972).

Sandford, D. S., and Chisholm, K. A.: "The Treatment of Municipal Wastewater Using the ICI Deep-Shaft Process," paper presented at the 29th Western Canada Water & Sewage Treatment Conference, Edmonton, Alberta, 1977.

Scaramelli, A. B., and DiGiano, F. A.: "Upgrading the Activated Sludge System by Addition of Powdered Carbon," Water and Sewage Wks., 120, 9, 90 (1973).

Schmitz-Lenders, F.: "Einige Probleme um unser Abwasser," Bauwirtshaft (Ger.), 39 (1954).

Schwartz, R. D., and McCoy, C. J.: "Use of Fluid Catalytic Cracking in Activated Sludge Wastewater Treatment," Jour. Water Poll. Control Fed., 48, 274 (1976).

Setter, L. R.: "Modified Sewage Aeration," Sewage Works Jour., 15, 629 (1943), and 16, 878 (1944).

Setter, L. R. et al.: "Practical Applications of Modified Sewage Aeration," Sewage Works Jour., 17, 669 (1945).

Shindala, A., and Parker, J. E.: "Thermophilic Activated Sludge Process," Water and Wastes Eng., 7, 47 (1970).

Siddiqi, R. H. et al.: "Effect of the Stabilization Period on the Performance of the Contact Stabilization Process," Jour. Water Poll. Control Fed., 39, 1211 (1967).

Van Soest, R., and Das, G.: "Tertiary Treatment Quality for a Secondary System Utilizing the Zurn-Attisholz Process," paper presented at the 28th Purdue Industrial Waste Conference, Purdue University, Lafayette, Ind., 1973.

Spady, B., and Adams, A. D.: "Improved Activated Sludge Treatment with Carbon," paper presented at the 47th Conference of the Water Pollution Control Federation, Denver, Col., 1974.

Stenquist, R. J. et al.: "The Coupled Trickling Filter-Activated Sludge Process: Design and Performance," EPA-600/2-78-116 (1978).

Stukenberg, J. R.: "Biological-Chemical Wastewater Treatment," Jour. Water Poll. Control Fed., 43, 9, 1791 (1971).

Surucu, G. A. et al.: "Aerobic Thermophilic Treatment of High-Strength Wastewaters," Jour. Water Poll. Control Fed., 48, 689 (1976).

Sutton, P. M., and Jank, B. E.: "Design Consideration for Biological Carbon Removal — Nitrification Systems," paper presented at the Seminar on High Quality Effluents, Toronto, 1975.

Sutton, P. M. et al.: "Nitrogen Control: A Basis for Design with Activated Sludge Systems," paper presented at the IAWPR Conference on Nitrogen as a Water Pollutant, Copenhagen, Denmark, 1975.

Sutton, P. M. et al.: "Single Sludge Nitrogen Removal Systems," Research Report No. 88, Canada-Ontario Agreement, 1979(a).

Sutton, P. M. et al.: "Oxitron Fluidized Bed Wastewater Treatment System: Application to High Strength Industrial Wastewaters," paper presented at 34th Industrial Waste Conference, Purdue University, May, 1979(b).

Toerber, E. D. et al.: "Comparison of Completely Mixed and Plug-Flow Biological Systems," Jour. Water Poll. Control Fed., 46, 1995 (1974).

Torpey, W. N.: "Practical Results of Step Aeration," Sewage Wks. Jour., 20, 781 (1948).

Torpey, W. N., and Lang, M.: "Effects of Aeration Period on Modified Aeration," Jour. Proc. Am. Soc. Civ. Eng., SA3, 1681 (1958).

Ullrich, A. H., and Smith, M. W.: "The Biosorption Process of Sewage and Waste Treatment," Sewage and Ind. Wastes, 23, 1248 (1951).

Ullrich, A. H., and Smith, M. W.: "Operation Experience with Activated Sludge-Biosorption at Austin, Texas," Sewage and Ind. Wastes, 29, 400 (1957).

Unz, R. F., and Davis, J. A.: "Microbiology of Combined Chemical-Biological Treatment," Jour. Water Poll. Control Fed., 47, 185 (1975).

Weers, W. A., and Andrews, J. F.: "The Effect of Contacting Patterns on the Transient Response of Activated Sludge Systems," paper presented at the 47th Annual Conference of Water Pollution Control Federation, Denver, Col., October 1974.

Wuhrmann, K.: "Ergebrisse von Grossversuchen an hochbelasteten Belebtschlammanlagen und Tropfkorpern," Schweiz. Zeit. f. Hydrologie (Swiss), 15, 1, (1953).

Wuhrmann, K.: "Nitrogen Removal in Sewage Treatment Processes," Vehr. Int. Ver. Limnol., 15, 580 (1964).

Zuelke, H. J.: "Biological Wastewater Purification Based on Aerobic Thermophilic Organisms," Fortschr. Wasserchem. (Ger.), 11, 153 (1969).

Aeration in Activated Sludge Process

5.1 GENERAL

The purpose of aeration in the activated sludge process is to supply oxygen
for micro-organism respiration requirements, to maintain micro-organisms
in suspension and to mix properly the contents of the aeration reactors.
After reviewing the oxygen transfer operation (p. 5.2) various types of
aeration equipment are presented and discussed (p. 5.3-5.6), followed by a
review of aeration reactors and their hydrodynamic features (p. 5.7). Description of automation of the operation of aeration (p. 5.8) closes this
Chapter.

5.2 OXYGEN TRANSFER OPERATION

THEORIES OF AERATION

The first model of mass transfer from the gas phase to the liquid phase
was formulated by Whitman in 1923 as the so-called "two-film" theory.
This concept assumed the existence of a gas film on one side of the gas-
liquid interface, and a liquid film on the other side of the interface. These
films were considered to be the major elements of the transfer mechanisms
as it was postulated that the transport through them followed the low of
molecular diffusion expressed mathematically by the Fick's formula:

$$\frac{\partial M}{\partial t} = -D\frac{\partial C}{\partial X} \tag{5-1}$$

where: $\partial M/\partial t$ = rate of mass transfer in moles per unit of time, D = dif-
fusitivity in surface units per unit of time, and $\partial C/\partial X$ = concentration
gradient across the film (Lewis and Whitman, 1924).

The Whitman "two-film" theory was modified in 1935 by Higbie, who formulated a gas penetration theory allowing for a time-dependent concentration profile. Ippen et al. (1952) adapted the two-film theory for air oxygen transfer into water. This approach was substantially modernized by Danckwerts who proposed in 1941 a stochastic surface renewal gas-transfer theory (Danckwerts, 1970).

AERATION KINETICS

Aeration kinetics of aqueous systems can be expressed in terms of the overall oxygen transfer coefficient K_La, or in mass terms, as oxygenation capacity (OC) in pounds or grams of O_2 per unit of time. Most simply, aeration kinetics are presented by a general differential equation:

$$\frac{dC}{dt} = K_La (C_s - C) \tag{5-2}$$

where: C = concentration of dissolved oxygen in water, mg/L C_s = concentration of dissolved oxygen at saturation level, mg/L and K_La = overall oxygen transfer coefficient, hr^{-1}. After integration, this equation can be given in the following form:

$$K_La = 2.303 (t_1 - t_0)^{-1} \log (C_s - C_0)/(C_s - C_1) \tag{5-3}$$

where: C_0 and C_1 = concentration of dissolved oxygen in water at observation times t_0 and t_1.

Oxygen saturation of water depends on water temperature (Table 5-1), barometric (Table 5-2) and hydrostatic pressure and dissolved solids contents in water (Table 5-3). The influence of water salinity and temperature on oxygen saturation values, can be also expressed by a simplified formula developed by Gameson and Robertson (1955):

$$C_s = (475 - 2.65S)/(33.5 + t) \tag{5-4}$$

where: S = salinity in g/L and t = temperature in °C. For diffused-air aerators submerged in aeration tanks at some depth, a correction is made for the increased oxygen patial pressure due to submergence. As the average hydrostatic pressure increases the oxygen saturation concentration, a half of the aeration tank water level depth is normally assumed.

The oxygenation capacity (OC) of a system or an aeration unit, as defined by Kessener and Ribbius (1934), refers to the temperature of 10°C and the pressure of 760 mm Hg. Therefore, the relationships between K_La and OC can be presented in the form:

$$OC = 11.33 K_La (k_{10}/k_\tau)^{1/2} \tag{5-5}$$

TABLE 5-1

Oxygen Solubility in Pure Water at Barometric
Pressure of 760 mm Hg
(after Carpenter, 1966)

Temperature in °C	Saturation Level in mg/L O_2
0	14.6
2	13.8
4	13.1
6	12.4
8	11.8
10	11.3
12	10.8
14	10.3
16	9.9
18	9.5
20	9.1
22	8.7
24	8.4
26	8.1
28	7.8
30	7.5
32	7.3
34	7.1
36	6.8
38	6.6
40	6.4

where: k_{10} and k_τ = coefficients of oxygen diffusion into water at the temperature of 10°C and at the measurement temperature τ°C, respectively (Table 5-4), and 11.3 is the saturation level of oxygen in water in mg/L at 10°C and atmospheric pressure of 760 mm Hg. This way of expressing the aeration kinetics was generally used in the fundamental studies on aeration performed by Pasveer (1956).

FACTORS AFFECTING OXYGEN TRANSFER

The values of $K_L a$ and OC are affected by the physical and chemical characteristics of the given aeration system. Of particular importance is the presence of surface active agents or other specific organic substances. These factors, to some extent, can be covered by introducing coefficients α (alpha) and β (beta) into the original equation.

The α coefficient was defined by Eckenfelder et al. (1956) as the ratio of $K_L a$ for the given solution to the $K_L a$ of tap water.

$$\alpha = \frac{K_L a \text{ (solution)}}{K_L a \text{ (tap water)}} \tag{5-6}$$

The practical range of the α-values observed at various wastewater treatment plants is from 0.6 to 1.2. This coefficient is expected to increase or decrease during the course of biological treatment to approach almost unity for treated wastewater, assuming that the substances affecting the transfer rate are being removed in the biological process. Thus, it usually has a

TABLE 5-2

Influence of Elevation above Sea Level on Oxygen Solubility in Water
(after Carpenter, 1966)

Temperature °C	Elevation in ft. (m x 0.3048)			
	.0	1,000 (304.8)	2,000 (609.6)	3,000 (914.4)
0	14.6	14.1	13.6	13.2
2	13.8	13.3	12.9	12.4
6	12.4	12.0	11.6	11.2
10	11.3	10.9	10.5	10.2
14	10.3	9.9	9.6	9.3
18	9.5	9.2	8.9	8.6
20	9.1	8.8	8.5	8.2
24	8.4	8.1	7.7	7.6
30	7.5	7.2	7.0	6.8

TABLE 5-3
Influence of Salinity on Oxygen Solubility in Water
at Barometic Pressure of 760 mm Hg
(after Carpenter, 1966)

| Temperature | Salinity, percent | | | |
°C	0	4.0	8.0	12.0
0	14.6	13.9	13.2	12.5
2	13.8	13.2	12.5	11.9
6	12.4	11.8	11.3	10.8
10	11.3	10.8	10.3	9.3
14	10.3	9.9	9.4	9.0
18	9.5	9.1	8.7	8.3
20	9.1	8.7	8.3	8.0
24	8.4	8.1	7.7	7.4
30	7.5	7.2	7.0	6.7

TABLE 5-4
Oxygen Diffusion Coefficients for Various Temperatures
of Oxygenation Capacity Measurements

Temperature °C	$\sqrt{\dfrac{k_{10}}{k_\tau}}$	Temperature °C	$\sqrt{\dfrac{k_{10}}{k_\tau}}$	Temperature °C	$\sqrt{\dfrac{k_{10}}{k_\tau}}$
1	1.187	9	1.019	17	0.878
2	1.165	10	1.000	18	0.861
3	1.142	11	0.982	19	0.845
4	1.119	12	0.964	20	0.830
5	1.098	13	0.946	21	0.815
6	1.077	14	0.928	22	0.799
7	1.057	15	0.911	23	0.784
8	1.038	16	0.895	24	0.770

greater effect on the aeration in plug-type flow aeration tanks than in completely mixed units.

The coefficient β is defined (Knop et al., 1964) as the ratio of the oxygen saturation for the wastewater concerned to the oxygen saturation for tap water:

$$\beta = \frac{C_s \text{ (solution)}}{C_s \text{ (tap water)}} \qquad (5-7)$$

After introduction of the coefficients α and β the basic aeration equation will be:

$$\frac{dC}{dt} = \alpha K_L a (\beta C_s - C) \qquad (5-8)$$

where $K_L a$ is the transfer coefficient for the air-water system under identical aeration conditions. Aeration kinetics obviously depend on aeration intensity, and for a compressed air aeration it can be expressed in the following way:

$$K_L a = aG_n^b \qquad (5-9)$$

where: G_n = rate of air flow, and a and b = constants. The α-values for many wastewaters are not constants, but depend on aeration intensity and the type of aeration device. Generally, lower α-values may be expected at a low energy input per aeration tank unit volume, and higher α-values at high energy inputs. Otoski et al. (1978) presented a comprehensive bench and full-scale study on the variability of alpha and beta coefficients of aeration.

It was also observed that the presence of micro-organisms in liquors under aeration enhances the aeration kinetics (α-values > 1). Originally, it was explained (Tsao, 1968) by the assumed direct absorption of oxygen from the atmosphere by the cells. Later on, Micka et al. (1973) suggested that it might be due to the physical presence of the suspension, probably because of the change in the nature of the micro-turbulence.

It is assumed that the aeration of the mixed liquor in an aeration tank may be represented by the formula:

$$\frac{dD}{dt} = -K_L a D + KL \qquad (5-10)$$

in which: D = oxygen saturation deficit; t = time; L = oxygen demand; and K = rate coefficient of oxygen demand.

AERATION PERFORMANCE TESTING

Aeration performance testing may be accomplished by aeration of deoxygenated tap water, sodium sulfite solution oxidation, biological oxygen uptake, and by tracer measurements (Neal and Tsivoglou, 1974). The most widely used test is based on the unsteady state aeration of deoxygenated water. The dissolved oxygen can be removed from water by stripping with nitrogen gas or, more commonly, by chemical reaction with sodium sulfite in the presence of cobalt catalyst. After starting aeration, the increase of dissolved oxygen in water is observed as a function of time. From these data, the oxygen transfer rate coefficient is calculated. In sodium sulfite oxidation, the reaction between sulfite ions and dissolved oxygen to form sulfate ions is utilized to determine the rate of oxygen transfer. However, this rate is strongly affected by the chemical reaction, and the respective results are not directly comparable with ones received by aeration of deoxygenated water. A thorough discussion of factors affecting oxygen transfer tests was presented by Boyle (1979).

The Ontario Ministry of the Environment (Smart, 1980) recommends, for the testing of aerators, a procedure close to one suggested by the American Society of Civil Engineers. The testing should be made in the full-scale aeration tanks, on chemically de-oxygenated tap water. For water deoxygenation, a sodium sulfite solution with cobalt chloride catalyst at a concentration of 0.25-0.5 mg/L should be used. The measurements of dissolved oxygen during the test should be made with several DO meters, located in selected points of the aeration tank and calibrated each day before testing. During the test run, the aeration tank dissolved oxygen concentration should be allowed to reach 90% of the saturation value. It is also recommended that the test results be processed graphically.

AERATION ECONOMY

Aeration systems and aeration devices can be rated in terms of their oxygen-transfer kinetics in pounds (or grams) of oxygen per horsepower-hour (or kW-hour) under standard conditions at zero dissolved oxygen contents. The values of aeration economy (called also aeration efficiency) may refer to the total energy requirement for aeration or may exclude energy losses not typical for the operation of oxygen transfer. The former values, which are of course numerically lower, are called "gross economy" of the operation, and the latter—the "net economy." For particular aeration systems, there are ranges of oxygen transfer rates for which aeration economy shows its optimal values.

STRIPPING EFFECTS OF AERATION

Intensive aeration and associated mixing in aeration tanks of the activated sludge process, produce conditions for possible stripping of volatile substances from the treated wastewater. In a study performed on simulated wastewater containing easily strippable and not highly biodegradable substances, Gaudy et al. (1968) demonstrated that an activated sludge treatment and air stripping proceeded simultaneously. In a later study, Tischler et al. (1978) found that under practical conditions, biological removal of organic substances usually is preferential to air stripping, and consequently, the effect of stripping may be very moderate even for hydrocarbons.

5.3 MECHANICAL AERATORS

Mechanical aerators are, generally the surface aerators which can be constructed as horizontal or vertical rotors. Aerators with horizontal axes include aeration brushes and aeration disks. Those with vertical axes are the low-speed or high-speed axial flow type. Oxygen transfer efficiency of mechanical aerators is usually expressed in pounds or grams (kilograms) of oxygen per brake horse-power and hr, at a zero oxygen content in the test aeration tank. Often referred to values of efficiency for such aerators are between 2 and 4 lb O_2/bhp-hr (0.7-1.3 kg 0_2/kWhr). In the past it was claimed that maintenance costs of mechanical aerators were higher than those of diffused air-aerators, as, usually, moving parts have to be changed more often. However, at present, due to the development of more reliable gear boxes and motors, the general economy of many types of mechanical aerators appears to be well comparable to diffused air aeration.

OXYDATION BRUSHES

Oxydation brushes as an aeration device (Fig. 5-1), were developed in Holland by J. N. H. Kessener in 1925-1927. Originally, the axes of these brushes were made of wood; they had a diameter of about 2 ft (60 cm) and rotated at the surface of the Kessener aeration tanks (see p. 5.7), close to one of its walls, at 50-70 rpm, at a relatively shallow immersion of a few mm. In 1932, Husmann patented a modification of Kessener brushes, composed of a number of thin stainless steel combs. A further development of this aeration system (Muskat, 1958; Baars and Muskat, 1959; von der Emde, 1964) produced different modifications of the device (Fig. 5-2) and was completed with the design of the so-called cage-rotor (Figs. 5-3 and 5-4).

Presently-used Kessener brushes have diameters in the range of 0.4 to 0.8 m. They are most frequently applied in oxidation ditches. The special mammoth rotors (Muskat, 1979) of a diameter of 1 m are used both in very large oxidation ditches and in the circulation tanks designed for larger wastewater flows (see p. 5.7).

AERATION DISKS

Aeration disks for orbal aeration systems (see p. 5.7) are perforated horizontal axes. The disks' sizes are 4 ft (1.2 m) or 4.5 ft (1.35 m) in diam-

Figure 5-1. Kessener brush of the cage-rotor type installed in Southhampton, Ontario (photo courtesy of P. Laughton).

eter. The larger disk is 0.5 in (1.27 cm) thick with an outer 13 in (33 cm) border perforated with approximately 3,350 holes of 0.5 in (1.27 cm) in diameter. The oxygenation efficiencies determined for these disks range from 1,820 to 2,000 g O_2/kWhr at 52 rpm at 18 in (46 cm) immersion.

LOW SPEED AERATORS

Low speed aerators are essentially low head, high volume pumps. Their impellers vary in design and size as well as in rotational speed and the power input. The reported diameters of these aerators are in the range from 3 to 12 ft (1-4 m), and they rotate at 30 to 85 rpm. The respective power imput are usually in the range from 3 to 150 HP (2.25 to 112.5 kW). Some of these aerators are equipped with draft tubes. They can be mounted on a bridge or pier (Figs. 5-5 and 5-6) or they can be designed as floating units (Fig. 5-7). Their impellers may be fabricated from steel, cast iron, various alloys and fibreglass-reinforced plastics. Generally, the oxygenation capacity of particular surface aerators increases with the diameter of the unit, to some number of rotations per minute and in the case of some of them with the immersion of the impeller. However, aeration economy for these aerators shows characteristic optima of rotational speed and immersion. There are numerous aerators of this type currently used and marketed. Some of them are described here as examples of various design possibilities.

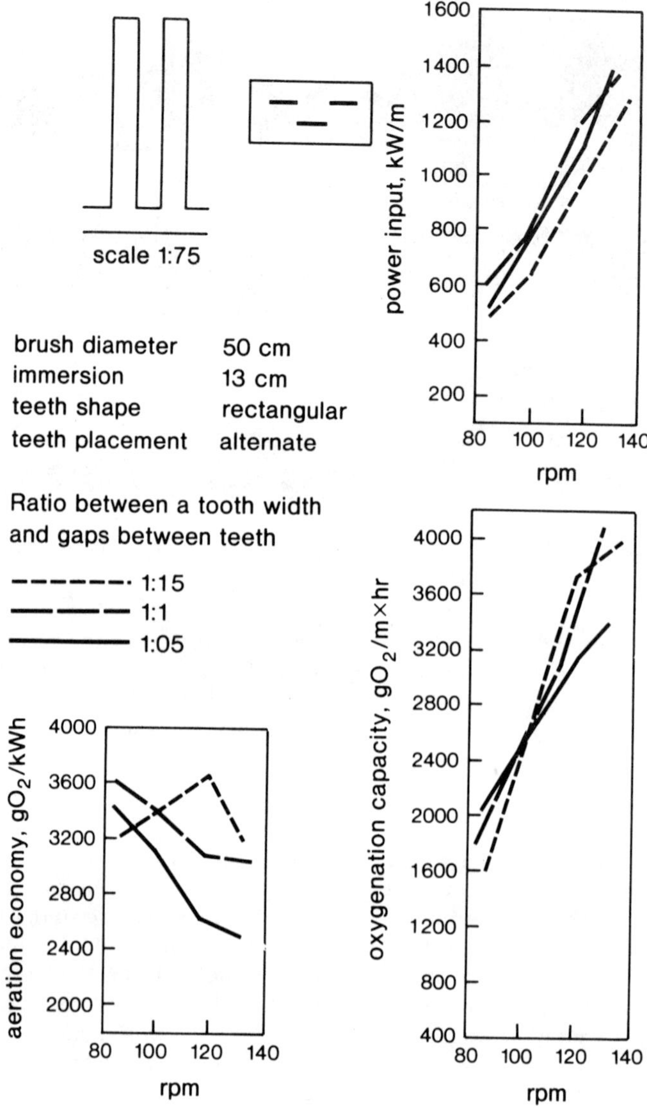

Figure 5-2. Operational characteristics of a modern Kessener brush (after Baars and Muskat, 1959).

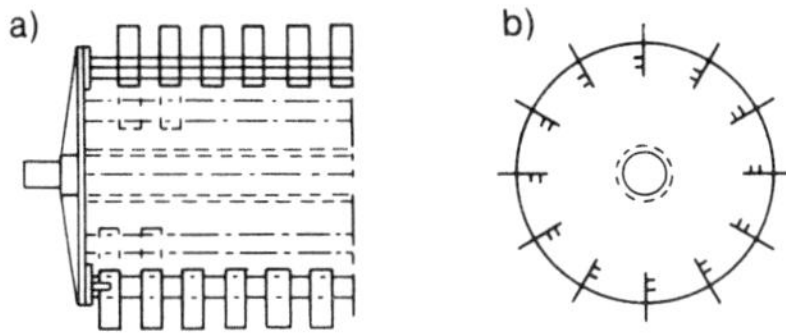

Figure 5-3. Cage rotor—A modification of a Kessener brush.

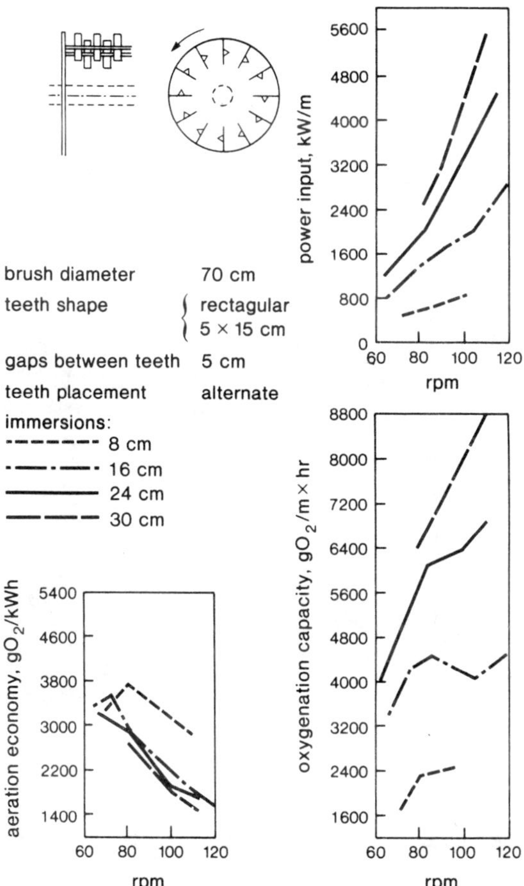

Figure 5-4. Operational characteristics of a Cage rotor (after Baars and Muskat, 1959).

Figure 5-5. Surface mechanical aerators mounted on a bridge in the Milton, Ontario Sewage Treatment Plant (photo courtesy of P. Laughton).

Vortair Aerators (Fig. 5-8) are relatively simple units, and now can be considered somewhat obsolete. They were the subject of relatively early studies by Weston (1962) who showed that a very attractive aeration economy could be achieved with the use of these facilities.

Lightnin Surface Aerators (Fig. 5-9) are available in two types: (a) with water level insensitive axial flow impeller, and (b) with level sensitive radial flow impeller, composed of 8-12 mild steel blades. The latter units are also characterized by limited splashing, and are quite popular in North America.

Simcar Aerators consist of a mild-steel cone with 130° angle, fitted with 8 vanes of mild-steel flats welded on edge to the underside of the cone and arranged tangentially from the hub situated at the apex of the cone

Figure 5-6. Surface mechanical aerators mounted on piers in the Dresden, Ontario Sewage Treatment Plant (photo courtesy of P. Laughton).

Figure 5-7. Floating aerator (photo by the author).

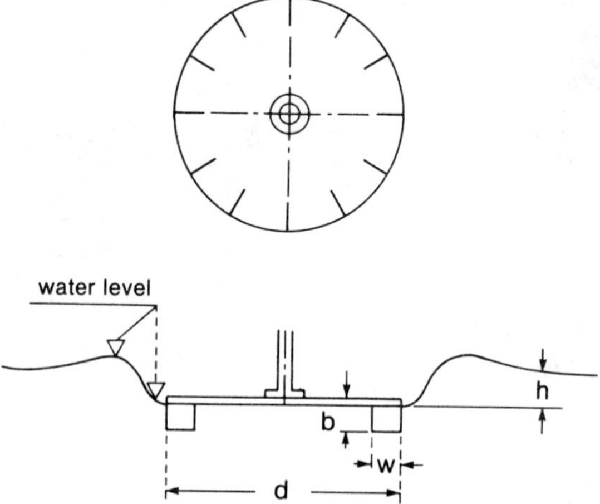

Figure 5-8. Vortair aerator (d, diameter; b and w, impeller size; h, immersion).

(Fig. 5-10). To enable the intensity of aeration to be adjusted over a certain range, a mechanical lifting and lowering device may be incorporated into each unit. Some results of a pilot-plant study of these units are presented in Table 5-5 (Knop et al., 1965).
Simplex Aerators were developed as early as 1916 by J. F. Bolton and are also called Bolton aerators. This system was studied very thoroughly by Downing et al. (1960). Various modifications of this type of surface aerator are still very popular.
BSK-Turbines (Fig. 5-11) are a very efficient surface aerator (Kaetin, 1964, and Stalmann, 1965), fabricated usually of light polyester resins. The optimal rotational speed is about 4.5 m/sec, and they are produced in diameters from 500 to 3,000 mm. Table 5-6 presents operational measurements of a BSK-turbine of 125 cm diameter, according to a study by Stalmann (1965).

MOTORS AND GEAR DRIVES

Most of the mechanical aerators for the activated sludge process are installed in aeration tanks on fixed platforms connected with proper walkways. This makes the inspection and maintenance of these units much easier. Usually, totally enclosed fan-cooled electric motors are used for these types of aerators. In the design of such motors there is no exchange of air between the outside atmosphere and the inside of the motor. This is especially advisable when using the motors for industrial wastewater treatment at plants where the formation of a corrosive atmosphere is likely. Under some circumstances, open drip-proof motors shielded from rain and mixed liquor spray, but with a free exchange of air inside the motors, may be used for sewage treatment. Aerator motors in the U.S.A. are mostly standard 460 volts units adapted for the 480 volts industrial electrical power distribution lines. In Canada, 575 volt units are used mostly. The insulation of these motors may follow different insulated classes or their combination. The life of the motors is expected to be at least 50,000 hrs. but can easily surpass 100,000 hrs. with adequate load control.

To match the changing aeration requirements with mechanical surface aerators capacity, it is possible to use multi-speed (usually two-speed) electric motors or multiple gear reducers. The gear reducers and gear drives used in the early development of mechanical aerators were characterized by a relatively short life. The current development of these units increases their life to about 12 years. This longevity, however, decreases rapidly with continuous or periodic overloading. Some of the overloadings may be caused by excessive rain and wind, fluctuating flow rates or, even more, by propeller icing. Lubrication of gear trains and bearings is

Figure 5-9. Lightnin aerator with deflector disc (photo courtesy of Greey Lightnin).

TABLE 5-5
Operational Measurements of a 229 cm Diameter Simcar Aerator
(after Knop et al., 1965)

| | Aerator | | | |
rpm	Immersion in cm	Aeration Tank Volume, in cu. m	Aeration Capacity in g O_2/cu. m x hr	Aeration Economy (gross) in kg O_2/kWhr
36	0	115.9	173	2.27
36	5	114.1	146	2.27
36	10	112.3	116	2.33
36	15	110.4	85	2.31
41	0	115.9	278	2.28
41	5	114.1	240	2.29
41	10	112.3	204	2.10
41	15	110.4	168	2.31

Note: Tap water was used as aeration medium.

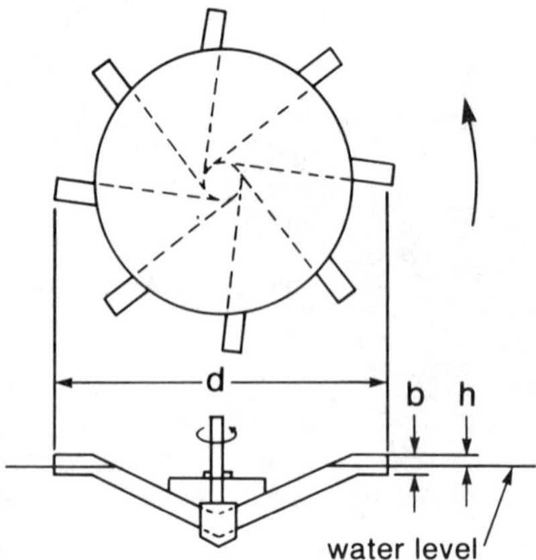

Figure 5-10. Simcar Aerator (d, diameter; b, flats depth; h, immersion).

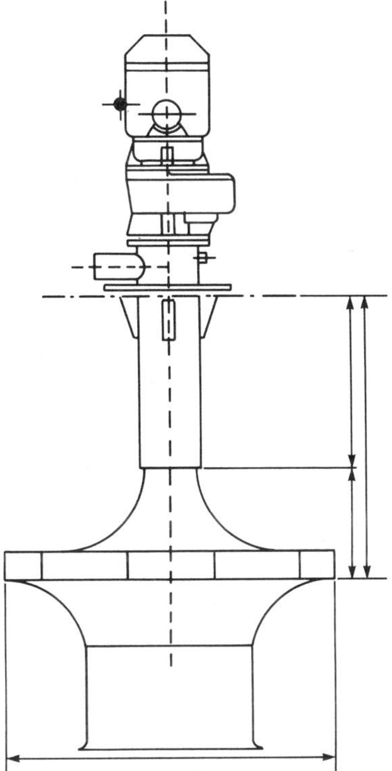

Figure 5-11. BSK-turbine aerator.

usually a positive lube type with either an oil pump driven by one of the
gear shafts or with oil bath splash lubrication. The oil change system for
these units should be designed in such a way as to prevent any accidental
spillings to the aeration tanks. It is advisable to enclose the drive-gears
in a cast-iron box to obtain the necessary corrosion protection. To protect the motors and gear drives of mechanical surface aerators against
overheating, overloading and vibrations, proper monitoring devices can be
installed temporarily or permanently. Usually, such steps are undertaken
only for relatively large units.

5.4 DIFFUSED-AIR AERATION

The introduction of compressed air to the aeration tanks in the form of
air-bubbles is the most common and most flexible oxygen transfer operation in the activated sludge process. Different types of diffusers are used

TABLE 5-6
Operational Measurements of a 125 cm Diameter BSK-Turbine Aerator
(after Stalman, 1965)

Aerator		Aeration Tank Volume, in cu. m	Aeration Capacity in g O_2/cu. m x hr	Aeration Economy in Kg O_2/kW hr	
rpm	Immersion in cm			Gross	Nett
60	0.0	52.7	76.5	2.16	3.84
60	2.5	53.3	98.7	2.39	4.15
60	5.0	53.9	92.4	2.24	3.68
60	7.5	54.5	97.7	2.27	3.43
60	10.0	55.1	102.9	2.18	4.88
65	2.5	53.3	118.5	2.57	4.21
70	0.0	52.7	139.5	3.03	4.88
70	2.5	53.3	150.3	3.08	4.86
70	5.0	53.9	158.1	2.79	4.25
75	2.5	53.3	152.9	2.59	3.85
80	2.5	53.3	176.5	2.68	3.93

Note: Tap water was used as aeration medium.

for this purpose; they produce fine, medium or coarse air-bubbles. Different placement of the diffusers affects the hydrodynamics of aeration reactors (see p. 5.7).

FINE BUBBLE DIFFUSERS

The application of fine bubble diffusers maximizes air utilization in the activated sludge process by maximizing the gas-liquid interface area per unit volume of air. The following media are offered for formation of fine air-bubbles: (a) vitreous-silicate-bonded grains of silica; (b) resin-bonded grains of silica; and (c) ceramically-bonded grains of fused crystaline aluminum oxide. These media are produced in the form of plates (Fig. 5-12), tubes (Fig. 5-13) or domes. Most popular are plate dimensions of 12 x 12 x 1-1/2 in (30.5 x 30.5 x 3.8 cm). Domes (raised disc-shaped diffusers) of 7 in (17.8 cm) diameter are mostly used. Recently, air diffusers made of porous polyethylene, like those made by the NOKIA Company

in Finland (represented in Canada by Greey Lightnin), have been developed.
These show a good oxygenation capacity and only require simple maintenance.

Most of the ceramic fine-bubble diffusers are characterized by permeability in the range from 15 to 21 units. Permeability is defined as the
rate of air flow, in cu. ft per min. at 70°F through a 1 sq. ft section of
diffuser of 1-1/2 in thickness, at a pressure drop equivalent to 2 in of
water column. The air flow rate should be corrected to standard conditions of 30 in Hg barometric pressure, 70°F, and 20 percent relative
humidity. Diffuser plates should have sufficient strength to permit uniform loading to 1,000 lb (454 kg) without breaking. Ceramic tube diffusers
are produced in different sizes. The most common type has a length of 24

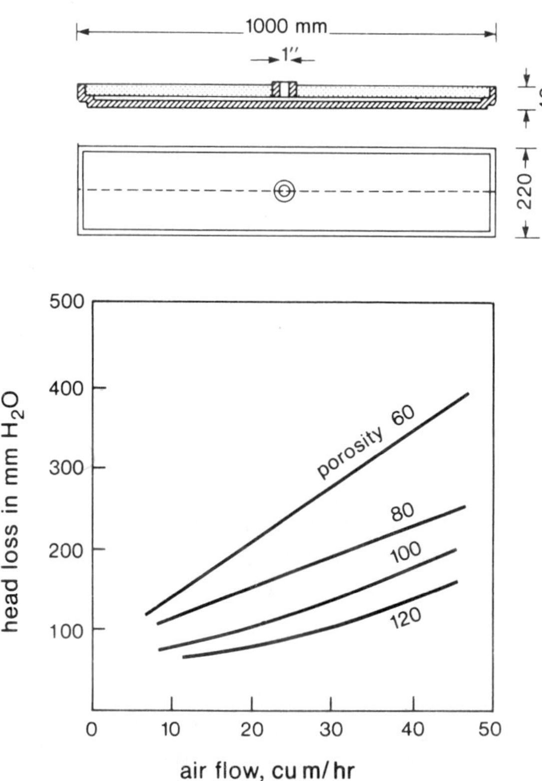

Figure 5-12. Brandol aeration tube and its characteristic head losses
(after Schumacher Works in Bietigheim, Wuerttenberg).

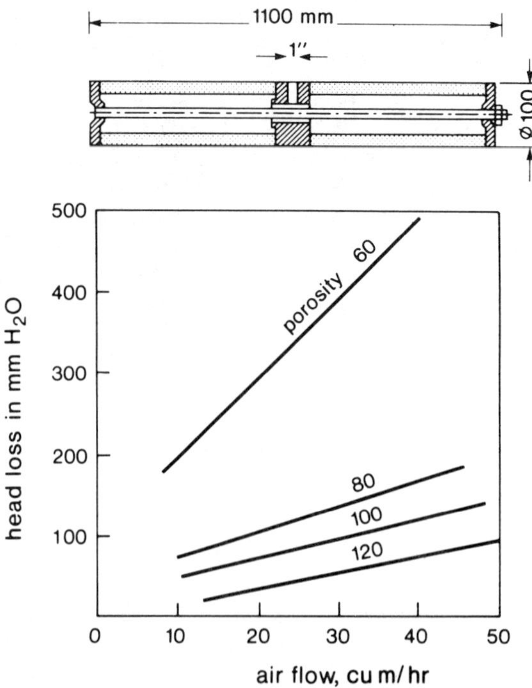

Figure 5-13. Brandol aeration tube and its characteristic head losses
(after Schumacher Works in Bietigheim, Wuerttenberg).

in, external diameter of 2-1/2 in and internal diameter of 1-3/4 in; its
nominal permeability is about 40.

Leary et al. (1969) examined seven different aeration systems in a
full-scale study to compare oxygenation efficiencies. The optimum was
found in the application of low-permeability diffuser plates. An example
of an aeration system composed of uniformly placed flat porous diffuser
plates may be the aeration tanks for the South Eastern Purification Plant
at Melbourne, Australia (Aberlay et al., 1974). During an extensive program of testing of the units, it was found that a uniformity of 'mixing' pre-
vailed and transfer efficiencies were as high as 13 to 19 percent at 15 ft
(4.6 m) submergence. Efficiency increased sharply at air flow rates of
40 to 50 cfm (1.13 to 1.42 cu m/min.).

MEDIUM-SIZE BUBBLE DIFFUSERS

Medium-sized air-bubbles can be generated by various porous diffusers.
The most common are perforated stainless steel tubes covered with a

spiral winding of saran cord, or covered by a sleeve of woven or knitted
plastic fabric (Fig. 5-14). These diffusers are considered to be relatively
expensive to maintain.

INKA AERATION SYSTEM

This aeration system, developed in Sweden (Fischerstrom, 1960), is based
on the introduction of low-pressure compressed air to the aeration tanks
through a grid of perforated pipes immersed only to about 2.6 ft (80 cm)
of the tank depth (Fig. 5-15). The Inka aeration grids cover up to one
half of the tank surface which is separated from the other part by a circulation wall reaching from the aeration grid to some depth over the tank
floor. The aeration economy of the Inka system depends on high-efficiency
air compression in special fans. Ganczarczyk (1964) studied the influence
of the design features of the Inka aeration grids, and their position in the
aeration tanks, on aeration kinetics and on the economy of this aeration
system.

COARSE BUBBLE DIFFUSERS

There is a multitude of devices developed to introduce air in the form of
coarse bubbles to the aeration tanks. Mostly, they are a metal or plastic

Figure 5-14. Medium-size air bubble diffusers (photo by the author).

Figure 5-15. Inka aeration system.

inserts attached to a submerged air header. They work as a nozzle, an
orifice, or a slit. Some of them have multiple openings, adjustable openings, and are equipped with a check-valve to prevent an intrusion of mixed
liquor in the event of interrupted air supply (Ganczarczyk and Suschka,
1962).

CLEANING OF CERAMIC DIFFUSERS

Depending on specific local conditions, the ceramic air-diffusers of an aeration system may require cleaning, to decrease air pressure head losses.
The major causes of diffuser clogging can be dust and scale loading from
inside and/or slime growth from outside. Before starting a diffuser clean-

ing operation, the cause of clogging should be determined. Diffuser outside
cleaning can be accomplished without diffuser removal or with diffuser removal. In the first case, sandblasting or oxygen-acetylene heat spalling of
the surface are most popular. Sometimes temporary improvements in the
diffuser operation can be achieved by physical removal of clogging material
by rapid increase of air flow, steam water hosing, scrubbing with detergents and solvents, etc. Cleaning of removed diffusers can be much more
thorough. Solvents like ammonia, sodium hydroxide, sulfuric acid, sulfuric acid and dichromate solution, nitrite acid, etc. are used. An application of ultrasonic vibration with solvent is very effective. This system
of cleaning can also remove the inside clogging.

ADDITIONAL EQUIPMENT

Additional equipment in compressed air aeration systems in activated sludge
wastewater treatment plants consists of air intake and distribution facilities,
air filtration facilities (mostly for fine bubble aeration plant only), and air
compression facilities.

AIR INTAKE AND DISTRIBUTION SYSTEMS

Air intake for activated sludge aeration should be located above grade level
to eliminate the input of ground dirt and snow during winter. It should be
large enough to reduce the inlet air velocity below a vacuuming effect, and
should also be protected against atmospheric precipitation. In the aeration
tanks, the air diffusers are attached either to one continuous air header
running through the entire length of the tank, or to a series of shorter air
headers that are supplied with air by vertical pipes. The latter construction can be designed in such a way as to allow particular segments of the
diffusers to be removed (e.g., by an application of "swing joints" on the
vertical header pipes). Such a design has distinct operational advantages.

AIR FILTRATION

Air filtration is required for fine bubble aeration, and is also advisable for
medium-sized and coarse bubble aeration. In the first case, an application
of unfiltered air would soon cause diffuser clogging; in the latter, the air
filtration also extends the life of air compressors. A different degree of
filtration is suggested for each of the above.

   Atmospheric dust contains particles from greater than 10μ to submicron
sizes. Often these particles have a tendency to agglomerate and thus can
plug passages larger than their size. The requirements for fine bubble diffusers call for air cleaning to levels lower than 0.05 to 0.09 mg of particulate matter per 1,000 cu. ft of air (0.02 to 0.03 mg/cu. m).

   There are three basic types of filters for air cleaning: (a) viscous
impingement filters; (b) dry barrier filters; and (c) electrostatic filters.
In viscous impingement filtration, the dust particles are contacted with

the coated surface of the filter and become entrapped until the filter is
cleaned or re-coated. In dry-barrier filtration, the dust particles are
retained on fine and closely packed media. In the electro-static filters,
the fine dust particles agglomerate and precipitate by applying a strong
electric field.

The filtration systems for fine bubble aeration often utilize more than
one type of filtration, e.g., electrostatic precipitation, followed by dry-
barrier filters. Another possibility is application of powder-coated cloth
bags attached to the shaking mechanism. The coat improves the filtration
effectiveness, and the shaking allows for a self-cleaning operation. Air
filtration for coarse bubble aeration usually employs the phenomenon of
viscous impingement of dust particles. The air filtration system for an
activated sludge treatment plant should be compartmentalized to ensure a
continuous operation, even in periods of maintenance cleaning or exchang-
ing of disposable parts.

AIR COMPRESSION

Many different devices are used for air compression at activated sludge
treatment plants. Their selection depends on technological requirements
and economic evaluation for the application at the specific plant. Table 5-7
presents a range of blowers available for activated sludge treatment needs.
Polytropic efficiency data for these blowers are given in Table 5-8. According to Szczesny (Gibbon, 1974) there is a trend towards the application
of the single stage centrifugal blowers because of their initial lower costs
and an efficiency comparable to the multi-stage centrifugal blowers. They
also produce less noise than other blowers. However, in the case of fine
bubble aeration, the growing headlosses on diffusers may reduce air output
from centrifugal blowers. On the other hand, under such conditions, the
positive displacement blowers would have just about the same output but
would require more power to operate.

5.5 COMMERCIAL OXYGEN AERATION

Commercial oxygen is used for activated sludge aeration in covered aeration tanks, in fluidized-bed reactors and in some types of open aeration
tanks.
Oxygen aeration in covered tanks is mostly connected with an application of
the specific modification of the activated sludge process like the UNOX system (see Chapter IV). Usually, various types of surface aerators are used
in the enclosed atmosphere over the mixed liquor level which is enriched in
oxygen (Fig. 5-16). The same approach may be used also for upgrading
some of the existing mechanical aeration plants. The largest European
activated sludge plant in Emschermuendung, West Germany, is equipped

TABLE 5-7
Available Blowers for Air Compression
(after Gibbon, 1974)

Volume Range	Blower-Types Available	Remarks
To 15,000 CFM (400 cu. m/min)	Lobe-positive displacement Modularized vertically-split Multi-stage centrifugal	Low first cost
15,000 CFM to 47,000 CFM (400-1,300 cu. m/min)	Integral-gear, single stage Pedestal-type single stage centrifugal Multi-stage, horizontally-split centrifugal	Lowest cost, minimum space Intermediate cost, more space Traditional approach, more costly (need extra stages to go direct drive)
47,000 CFM to 100,000 CFM (1,300-2,800 cu.	Pedestal-type single stage centrifugal Multi-stage, horizontally-split centrifugal—single or double inlet	Lowest first cost at comparable efficiency Traditional approach, more costly
100,000 CFM to 150,000 CFM (2,800-4,200 cu. m/min)	Pedestal-type, single stage centrifugal	Lowest first cost
100,000 CFM to 200,000 CFM + (2,800-5,600 + cu. m/min)	Axial	Highest cost—highest efficiency. Evaluates best over long operation period

with Simplex aerators which, due to a different geometry of aeration tanks than those used in pre-design studies, are not supplying enough oxygen. In consequence, the plant has an odour problem. To solve this problem, it is intended to cover the tanks and enrich the atmosphere below the covers in oxygen by introducing commercial oxygen. Also the idea of the pipeline activated sludge process (U.S. Patent 3,607,735) could be considered as

TABLE 5-8
Polytropic Efficiency of Blowers
(after Gibbon, 1974)

Blowers	Efficiency in %
Lobe	71%-78%*
Modularized, vertically split multi-stage centrifugal	71%-77%
Single stage centrifugal (integral gear)	76%-79%
Single stage centrifugal pedestal type (separate gear)	77%-80%
Multi-stage centrifugal	78%-80%
Multi-stage axial	82%-83%

*For low flows at approximately 2000 CFM at
8 PSI. (60 cu. m/min at 0.6 kg/sq. cm).

another possible application of oxygen activated sludge. In this design, the
mixed liquor is pumped into a pipeline reactor where it is contacted with
commercial oxygen introduced in several different ways. Of course, such
treatment has to be equipped with a secondary clarifier and a recyle system.
Use of such an approach could be suggested where a wastewater collection
system calls for a design which could be extended into such treatment, or
where land use is extremely restricted.

OXYGEN AERATION OF FLUIDIZED BEDS

The extremely high oxygen requirements of the fluidized-bed reactors (see
Chapter IV) call for very high oxygenation capacities of the applied "aerators". Therefore, commercial oxygen, rather than air, should be used
for this purpose, and techniques used for solubilization of oxygen should be
very efficient. One possible solution is the application of the patented
Dorr-Oliver contactor or of U-tubes for such an oxygenation (Speece,
1979; and Speece et al. 1980). These units (Fig. 5-17) consist of two concentric pipes which extend for about 200 to 300 ft. (61 to 92 m) into the
ground. Commercial oxygen is injected into the inner pipe and travels
downward with the forced flow of water. The effective contact time and
the increased hydrostatic pressure resulting from the depth of the U-tube
allow high oxygen transfer efficiency, high resulting DO concentration, and

cost-effective economics. According to Speece (1979), the approximate
optimum operational conditions for such units are 75 m depth, 1.5 m/s
throughput velocity and 66 mg/L outlet DO concentrations with 100% oxygen
absorption efficiency. Commercial oxygen currently (1980) costs between
$20 and $100 per ton, depending on quantity used and on transportation
costs or production on the site.

AERATION IN OPEN TANKS

There is also a possibility of application of oxygen aeration in a normal or
somewhat modified open aeration tank. One such system, called MAROX
(Stetzer, 1975) utilizes hydraulic shear of oxygen gas to produce micron
size gas bubbles from a rotating diffuser. This system can be installed in
existing aeration tanks to upgrade their performance. A full-scale demonstration study for this system was performed in Denver, Colorado (Pearlman and Fullerton, 1979).

Another open tank oxygen aeration system is called <u>forced free fall</u>

Figure 5-16. Covered aeration tanks equipped with surface aerators for
Oxygen aeration at the UNOX Treatment Plant in Winnipeg, Manitoba
(photo courtesy of Union Carbide).

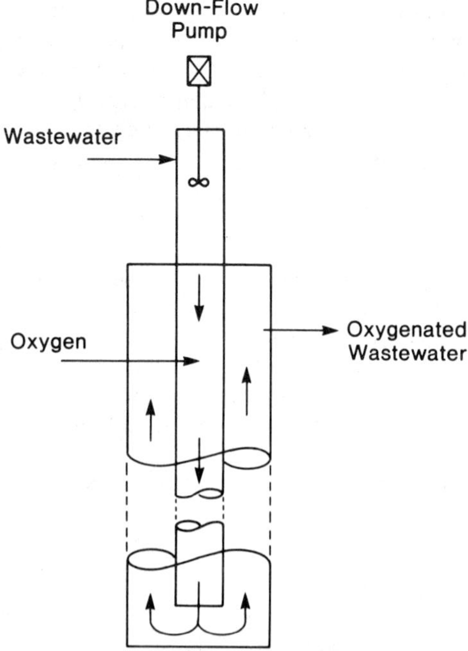

Figure 5-17. U-tube oxygen contactor.

<u>oxygenation</u>. It has been developed by Airco Incorporated (1975). This
system provides commercial oxygen dissolution in dome modules immersed in aeration tanks. Mixed liquor flows into a module, and because
of a pump action goes through a waterfall in an oxygen-rich gas space.
Some retention in a high turbulence fall zone and later in a quiescent zone
improves the oxygen dissolution and recovery of undissolved gas. The
oxygenated mixed liquor flows through a nozzle at the base of the module
back to the aeration tank at a velocity sufficient to maintain mixing and
sludge suspension in the tank. The control of this oxygenation system is
through changes in oxygen concentration in the gas zone of the module.

OXYGEN SUPPLY

Oxygen supply for the oxygen activated sludge plants can be designed as an
on-site oxygen generating plant, or as an oxygen supply from another oxygen generating plant with only liquid oxygen storage and vaporization facili-

ties on site. Oxygen generation (McWhirter, 1978) can be accomplished with the use of the cryogenic air-separation method applicable for large installations or by the pressure-swing adsorption system (PSA) where, somehow, smaller quantities of oxygen are required (Fig. 5-18). The cryogenic air-separation method involves the liquefaction of air, followed by

Figure 5-18. PSA oxygen generation in the Tucson, Arizona, Ana Road Wastewater Treatment Plant (photo courtesy of P. Laughton).

fractional distillation to separate nitrogen and oxygen. The pressure-swing adsorption system uses multibed adsorption to provide a continuous flow of oxygen. The feed air is compressed and passed through one of the adsorbers which removes carbon dioxide, nitrogen and humidity from it. Simultaneously, other adsorbers are in various stages of regeneration. Power requirements for oxygen generation are in the range of 13 to 17 kW per ton of daily capacity (6.4 to 4.9 lb of oxygen produced per kWhr = 2.2 - 2.9 kg/kWhr).

5.6 OTHER AERATION METHODS

Combining air diffusion aeration with some kind of mechanical mixing and/or surface aeration is the basis of such aeration methods as aeration turbines. Similarly, some enhancing of diffused air aeration can be accomplished by the application of static tubes and various jet aerators. Recently, the development of reversed draft tubes produced two new and interesting aeration systems: the draft tube aerator and the induction mechanical aerator. It is also possible to use a modification of biological filtration for mixed liquor aeration (Spirovortex system), and to use chemically combined oxygen for activated sludge respiration.

<u>Submerged turbine aeration systems</u> are based on compressed air introduction through open pipes or perforated sparge rings, located below one or two submerged impellers (Fig. 5-19 and 5-20), mounted on a vertical axis. The air bubbles formed are broken into small fragments by moving impeller blades, and the contents of the aeration tanks are simultaneously mixed by the impeller rotation. The aeration effects with turbine aerators depend on air flow and on type, size and number of rotations of the impeller or impellers. Usually this information is supplied by the equipment manufacturer. In older designs the submerged turbine aeration was often associated with a mechanical surface aeration.

<u>Static type aerators</u> are based on an air bubble release under submerged vertically-mounted tubes containing some baffles (Ball et al., 1974, Hsu et al., 1975, and Stewart and Lidkea, 1976). This aeration system is mostly used in aeration lagoons but it is used also in some activated sludge aeration systems.

<u>Jet aeration</u> can be accomplished by mixing pressurized air and water within a jet nozzle and discharging the mixture into aeration tanks (Toerber and Mandt, 1979). Relatively high oxygen transfer rates are reported for such systems (l.c.). Another type of jet aeration is based on a surface application of water-jet (mixed liquor spray) from a perforated pipe rotating around a circular aeration tank (Jennekens, 1979).

<u>Down-flow draft tube aerator</u> consists of an air sparger located in draft tube beneath an impeller creating a down-flow movement of mixed liquor. Under these circumstances oxygen transfer is enhanced because of the

Figure 5-19. Submerged turbine aeration system (photo courtesy of Greey Lightnin).

Figure 5-20. A submerged turbine with an additional surface aerator (photo courtesy of Greey Lightnin).

increased time of air bubbles retention, and the increased hydrostatic
pressures at deeper installations. This type of aeration unit is recommended for the treatment of high-strength wastewater, when land area is
limited and aeration tanks must be deep, and for the draft tube channels
(see p. 5.7).
<u>Induction mechanical aerator</u> for aeration tanks and aerated lagoons, con-
sists of a mechanical drive with a hollow, low-speed shaft and a hollow
impeller within a draft tube. Air is induced (sucked) into the mixed liquor
through the hollow shaft by the low pressure developed at the rotating impeller, and is jetted through holes along the top of each hollow blade. A
fine dispersion of bubbles is then driven down through the draft tube to the
bottom of the tank.

AERATION IN BIOLOGICAL FILTERS

The Spirovortex system (Nelson, 1958) applies a special high-rate, coarse-
media biological filter for aeration of mixed liquor. Similarly an Activated
Bio-Filter system (Owen and Slechta, 1975) uses a flow of recycling mixed
liquor through a biological filter (called bio-cell) consisting of evenly
spaced, horizontal redwood slats (see also Chapter IV, p. 4.8).

USE OF CHEMICAL OXYGEN

For activated sludge treatment of some industrial effluents containing a
substantial concentration of nitrate, utilization of this form of oxygen may
be an attractive solution to aeration needs, combined with nitrogen elimination (McKinney and Conway, 1957; Miyaji and Kato, 1975, etc.). Such
treatment is virtually a denitrification process (see Chapter IV, p. 4.9).

5.7 AERATION REACTORS

Biological reactors for the activated sludge process are called aeration
tanks because one of their main features is to supply oxygen for the microbial transformations. The hydrodynamic characteristics of aeration tanks
can be determined by specification of resulting mixing (see Chapter III,
p. 3.6) and a minimal bottom horizontal flow velocity. In addition, the
turndown and, for some mechanical aerators, the draft tubes (if any) circulation, may also characterize the hydraulic regime of these units.
There is general agreement that velocities of at least 0.5 feet per second
(15 cm/sec) are required to keep biological floc in suspension. The rate
of the hydraulic turndown of mixed liquor in the aeration tanks may be
determined as the turnover-time defined as the tank volume divided by
pumpage rate. Therefore, although tank geometry affects the complex
mixing pattern in the tank, it may be assumed that distance travelled by

an increment of pumped flow is tank diameter plus twice the water depth.
Maximum effective pumping rate of mixed liquor by release of compressed
air at a submergence of 15 feet (4.5 m) of water can be estimated as 7.2 cfs
(0.2 cu. m/sec) per horsepower input. The parameter of power input per
unit volume (e.g., HP/1,000 cf) may be also used to roughly describe the
required turbulence in aeration tanks.

The aeration tanks can be divided according to their physical shape
into (1) retangular tanks, (2) circular tanks, (3) circuit tanks, (4) tanks of
special shape, and (5) aeration tanks combined (or integrated) with secondary clarifiers. All these aeration tanks can be equipped with various
aeration systems, described in p. 5.3-5.6.

Rectangular aeration tanks with a water level depth of 10-15 ft (3-4.5 m)
and a freeboard of up to 5 ft (1.5 m) to contain spray from surface aerators
and/or foam, are the most commonly used tanks for activated sludge
aeration. Normally they are constructed of steel or reinforced concrete
and left open to the atmosphere. Sometimes, however, they are covered
for air pollution or climatic reasons or even enclosed for a surface diffusion of commercial oxygen. Individual mechanical surface aerators located in the same aeration tank should be installed far enough apart to
avoid hydrodynamic interference with one another. Such interactions are
called "surging". Surging can be prevented by installing baffles of proper
sizes and at proper locations. The rectangular aeration tanks aerated
with a diffused air aeration system can be divided into ridge-and-furrow
tanks, spiral flow tanks, Inka aeration tanks, etc. (Fig. 5-21. The
"fixed activated sludge" tanks were described in Chapter III, p. 3.5.

Ridge-and-furrow tanks. The bottom of the ridge-and-furrow tanks, called
also Milwaukee tanks, is uniformly covered with air diffusers (Fig. 5-21).
When first introduced, these tanks had aerators in furrows to prevent
sludge deposition. However, this was found to be unnecessary except for
widely spaced aerators. Modern tanks are flat bottomed with a high density of aerators.

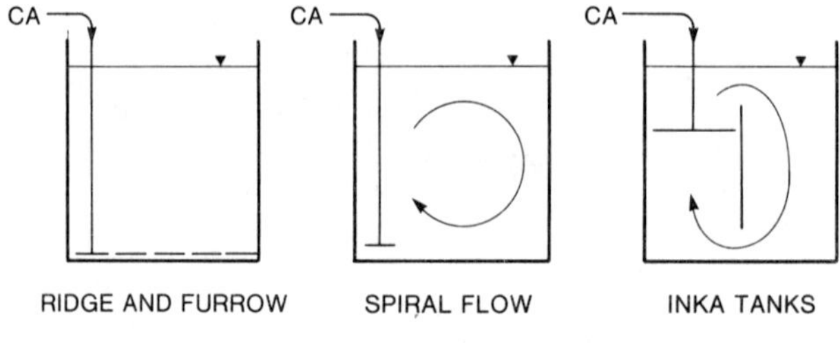

Figure 5-21. Schematic drawings of rectangular, diffused-air aeration
tanks.

<u>Spiral-flow tanks</u>. In the spiral-flow tanks (Fig. 5-21) air diffusers are placed along one side at the bottom of the aeration tank throughout its entire length (Hurd, 1923). Such an arrangement of diffusers causes a helical flow pattern of the mixed liquor under its longitudinal movement. This flow pattern is commonly but inaccurately referred to as a spiral flow. By placing diffusers in the centre of the tank floor or at either wall, it is possible to achieve a spiral flow with a double roll. The first modification may lead to sludge deposition problems but gives a better oxygenation capacity than the second one.

<u>Inka aeration tanks</u> (Fig. 5-15 and 5-21) are characterized by the typical air distribution grid (see p. 5.4) immersed usually at 2.6 ft (0.8 m) below the water level and the proper circulation baffle. Air flow through the distribution grid acts as an air pump and facilitates an intensive circulation of mixed liquor.

<u>Deep aeration tanks</u>. Interest in the design of deep aeration tanks is mostly the function of the limited space availability. An example of such units may be the design described by Rose and Gorringe (1974), in which each steel aeration tank had a diameter of 120 ft (36.6 m) and 24 ft (7.3 m) water depth. Aeration and mixing was accomplished by two aerators (closely spaced spargers in a rectangular draft tube) in each tank. The air was released at a submergence of 12.5 ft (4.1 m). The draft tube started at 5.0 ft (1.64 m) above the bottom of the tank and ended at 2.0 ft (0.66 m) above the point of air release. The vertical current induced by the rising air was deflected by the surface energy recovery baffle and directed horizontally across the surface. This action very effectively rolled the tank content and maintained thorough mixing and highly efficient oxygenation.

<u>Covered and enclosed aeration tanks</u>. To minimize air pollution in some underground or housed sewage treatment plants, it is the practice in Japan to cover aeration tanks with light corrugated plastic sheets, and direct the exhaust air to ventilation ducts. In turn, the exhaust may go through hypochlorite washers and/or be treated with ozone. For aeration tanks exposed to the atmosphere, covers are usually made of concrete or aluminum. The same is valid for tanks enclosed for the purpose of aeration with commercial oxygen (Fig. 5-16). The "Bayer Tower Biology R " can also be considered as an example of an enclosed aeration tank; it is a 98 ft (30 m) tall enclosed steel tank used for activated sludge treatment of industrial effluents from Bayer A G in Leverkusen, West Germany (Schnellman, 1980).

<u>Circular aeration tanks</u>. Separate circular aeration tanks are rarely used although a proper system of baffles may adapt them to many practical aeration systems. However, circular aeration channels are quite common in smaller plants which were designed as structurally connected units. Such channels may be aerated with diffused-air aerators or with mechanical aerators (Fig. 5-22).

<u>Circuit aeration tanks</u> have recently gained substantial popularity. The most popular are oxidation ditches (or channels) and orbal aeration tanks.

Figure 5-22. Old diffused air aeration system at Pickton, Ontario, converted into surface mechanical aeration (photo courtesy of Greey Lightnin).

<u>Oxidation ditches</u>. Oxidation ditches as activated sludge reactors are essentially closed-loop, open channels (Fig. 5-23) in which the mixed liquor (usually with a high suspended solids content) is aerated and circulated with the use of various devices. In the original design of oxidation ditches the mixed liquor aeration and circulation is accomplished by application of Kessener brushes or cage rotors (see 5.3). Other designs suggest for this an application of surface turbines called the Carrousel system and shown in Fig. 5-24 (Zepper and de Man, 1970), jet pumping of mixed liquor (Wong-Chong et al., 1974), reversed draft tube aeration (Fig. 5-25), etc.

Adequate mixing in the oxidation ditch to prevent sedimentation of the activated sludge is obtained by maintaining a flow velocity of about 1.0 ft per sec. (0.3 m per sec.). For lower levels of mixed liquor suspended solids, somewhat lower velocities may be satisfactory. Shallow ditches are usually 4-6 ft deep (1.22-1.83 m) with 45° sloping walls. Deep ditches have vertical side walls, and are normally 10-12 ft (3.05-3.66 m) deep.

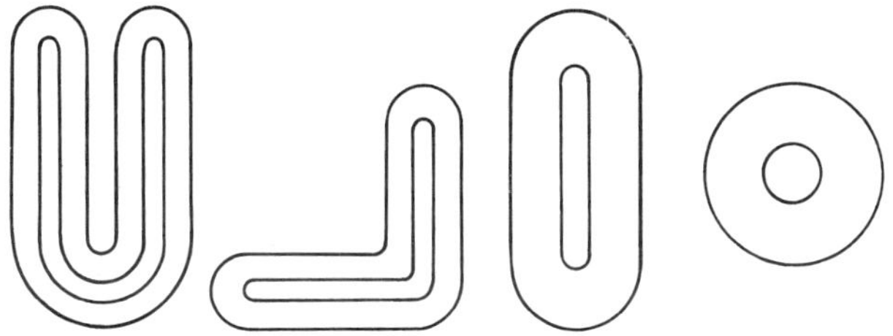

Figure 5-23. Typical configurations of oxydation ditches.

Ditches are often lined with reinforced concrete, gunite, asphalt, or thin plastic membranes to prevent erosion and leakage. Vertical walls of deep ditches normally require reinforced concrete.

For the secondary treatment of municipal wastewater, a comparison of oxidation ditch plants to competing processes (Ettlich, 1978) shows this approach to be economically attractive for plants under 10 mgd (37,850 m^3/day). Oxidation ditch design can also be applied to industrial wastewater treatment (e.g., Paulson and Lively, 1979).

Orbal Aeration Tanks (Fig. 5-26) are multicompartment, concentric chan-

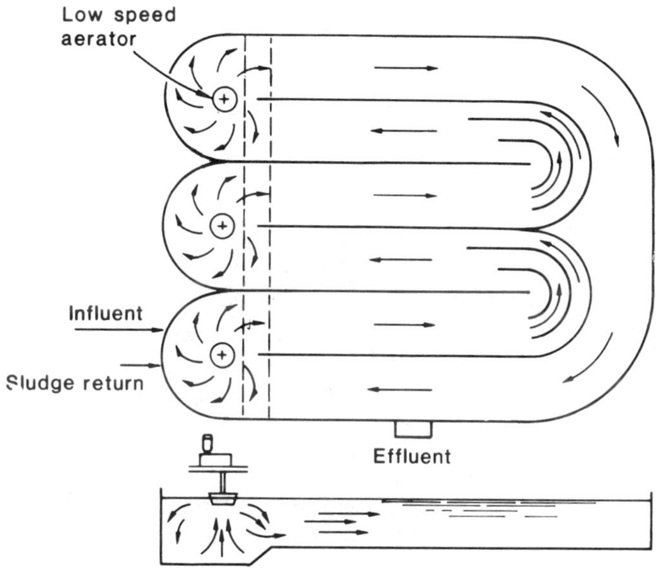

Figure 5-24. Carrousel system (after Envirotech Corporation).

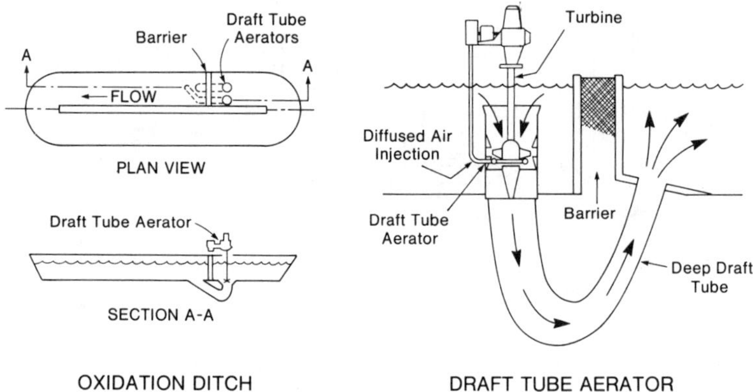

Figure 5-25. Oxidation ditch with draft tube aerator (after Lightnin).

nels forming a succession of orbal circuits, and equipped with disk aeration units (see p. 5.3). These channels are interconnected by ports at bottom level, so positioned that the outlet from any one channel is upstream of the corresponding inlet (Drews et al., 1972). Normally three or four channels are recommended for an activated sludge treatment system. Up to 1972, more than 50 orbal plants with various modifications were constructed in South Africa. These systems seem to be especially suitable for the extended aeration modification of the activated sludge process.

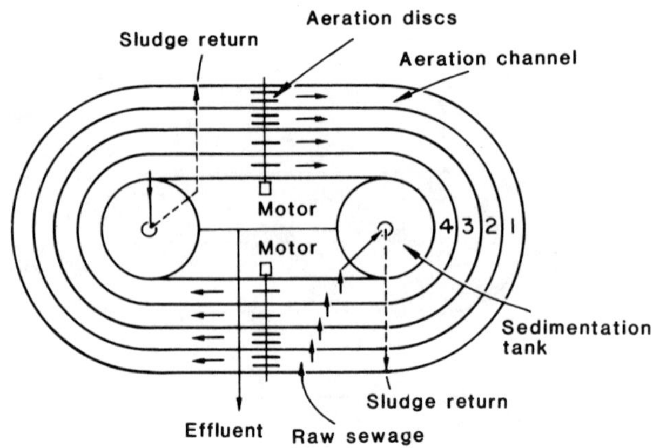

Figure 5-26. Orbal aeration system (after Drews).

<u>Aeration tanks of a special shape.</u> It is difficult to classify a number of
contemporary designs of aeration reactors in conventional terms for aeration tanks. They are "deep shafts", fluidized beds, multi-stage towers,
etc.

<u>Deep-shaft reactors.</u> Research into the production of single cell protein by
aerobic bacterial fermentation of methanol, carried out in Great Britain by
Imperial Chemical Industries (ICI), has brought about the development of a
novel aeration tank design (Bailey et al., 1975). This system operates as
a modified air-lift fermentor and is characterized by a substantial depth
(Fig. 5-27). It is suitable for mixed liquor aeration in a number of activated sludge process modifications and in aerobic digestion of organic
solids. As this system shows a very high oxygen transfer efficiency, its
application is suggested for treatment of raw sewage and/or concentrated
organic effluents. To start up the operation of the deep-shaft aeration
tank (Fig. 5-27) compressed air is initially injected at a comparatively
shallow depth into the part of the unit with the rising flow. This causes
mixed liquor circulation in the whole system. Once the circulation is

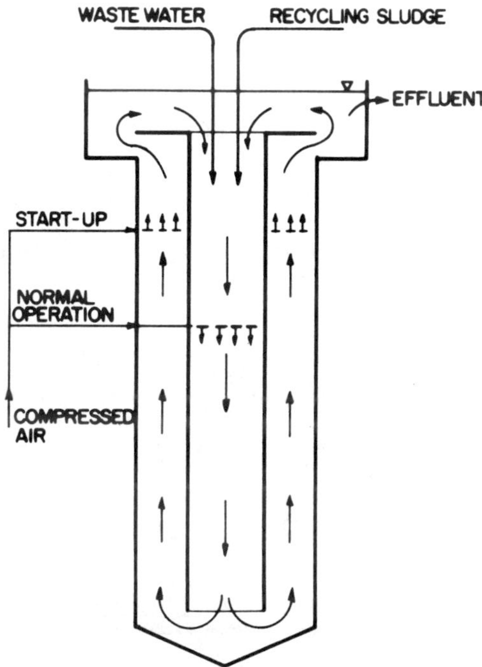

Figure 5-27. Deep-shaft reactor.

established, air injection is gradually transferred into the part of the
unit with the down flow. Continuous circulation is then supported by the
net change of gas voidage in both sections of the unit. Gas disengagement
is required at the head of the shaft.

The suggested depths of the shafts exceed 300 ft (100 m), and the expected transfer economy surpasses substantially the results achievable with
other systems of aeration. For practical reasons it is more convenient to
build these units rather as shafts than as towers. Depending on the required size and geological conditions the shafts can be mined or drilled.
It has been found that pressure fluctuations in this system do not affect the
bacterial sludge physiology. Moreover, some observations seem to indicate that under the applied conditions the carbon-to-cell conversion is somewhat lower than in conventional reactors (as it is claimed also for the Unox
process).

<u>Fluidized-bed reactors</u> are schematically shown in Fig. 4-7. The respec-
tive technology is presently being developed (see Chapter IV) and different
modifications of the reactor design are taken into consideration (Jeris
et al., 1977). Hermanowicz and Ganczarczyk (1981) studied hydrodynamic
behaviour of a three-phase (gas-liquid-solids) laboratory-scale fluidized-
bed model. It was found that both coefficients of liquid dispersion and
gas-to-liquid mass transfer were independent of the liquid flow rate, but
highly dependent on gas flow rate. Oxygen transfer efficiency was con-
strained due to gas bubbles coalescence, and to prevent slug formation,
porosity of the bed higher than some critical values was required.

<u>Multi-stage tower reactors</u>. Besik (1973) reported a succesful application
of a modified sieve-plate tower as a multi-stage, combined sludge activated
sludge aeration reactor. Each plate formed a separate mixing chamber.
The air passed upward in the tower, and the wastewater and the recycling
sludge passed downward from the upper plates to the lower ones. This
design was suggested for small water renovation installations, but its
practicability is somewhat questionable.

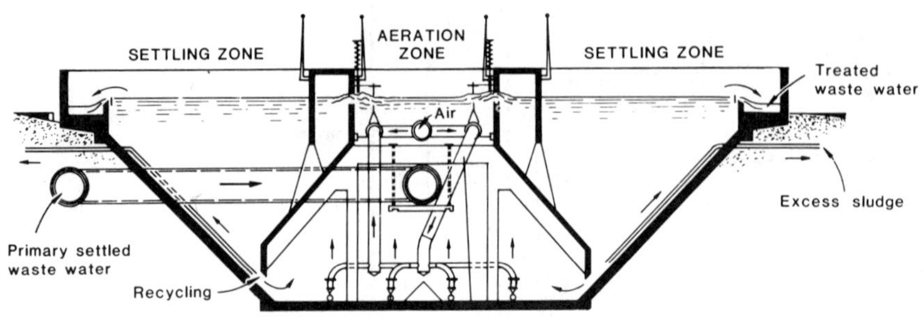

Figure 5-28. Degremont's "oxycontact" combined tanks.

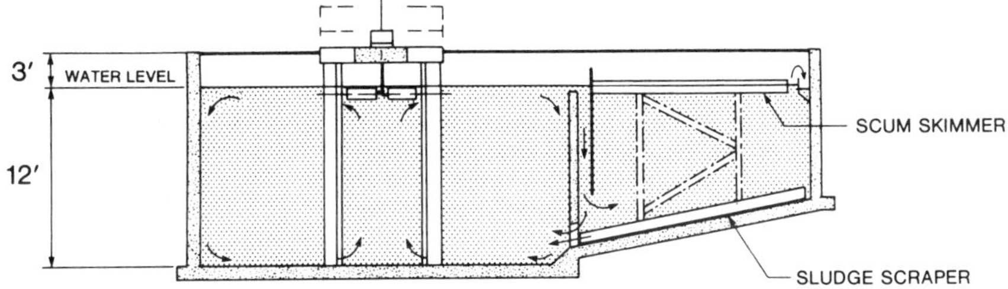

Figure 5-29.   Lightnin's integral clarifier system.

<u>Aeration tanks combined with secondary clarifiers</u>. Efforts to reduce the
construction costs of separate aeration tanks and secondary clarifiers as
well as the costs of construction and operation of a separate sludge recycle
system, have led to the development of various combined systems of these
two most important parts of the activated sludge process. The most common practical solutions are the Degremont Oxycontact system which is used
widely in Europe (Fig. 5-28) and the Lightnin Treatment system which is
popular in North America (Fig. 5-29). The Lightnin system is based on the
patent by L. S. Love (see Chapter VI, p. 6.2).

5.8 AUTOMATIC CONTROL OF AERATION

Automatic control of aeration in the activated sludge process (see Chapter X) improves the process performance and generally leads to meaningful energy savings (Roesler, 1974; Gonthe et al., 1978; etc.). The automation equipment for aeration control consists of a system of dissolved oxygen
sensors placed in critical areas of the aeration tanks, a dissolved oxygen
monitor, and a control unit connected with the applied aeration units (e.g.,
with air blowers for compressed air aeration or with particular mechanical
aerators). Signals from dissolved oxygen sensors are directed through the
controller to the regulating devices of the aeration system. Detailed design procedures for each control systems were presented by Flanagan and
Bracken (1977).

REFERENCES

Aberley, R. C. et al.: "Air Diffusion Unit," Jour. Water Poll. Control
Fed., <u>46</u>, 895 (1974).

Airco, Inc.: "Understanding the F^3O System," Murray Hill, N.J., 1975.

Baars, J. K., and Muskat, J.: "Zuurstoftoevoer an Water met Behulp van Roterende Lichamen," Instituut voor Gezondheitstechnik TNO (Holland), Raport No. 28 (1959).

Bailey, M. et al.: "Deep Shaft Aeration System," Engineering Digest (Canada), 22, No. 4, 28 (1976).

Ball, R. O. et al.: "Static Aeration Systems: Problems and Peformance," paper presented at the 29th Purdue Industrial Waste Conference, Purdue University, Lafayette, Indiana, 1974.

Besik, F.: "Multistage Tower-type Activated Sludge Process for Complete Treatment of Sewage," Water and Sewage Wks, 120, 9, 122 (1973).

Boyle, W. C. (editor): "Proceedings: Workshop Toward an Oxygen Transfer Standard," EPA-600/9-78-021, April 1979.

Carpenter, J. H.: "New Measurements of Oxygen Solubility in Pure and Natural Waters," Limnol. Oceanog., 11, 264 (1966).

Dankwerts, P. V.: "Gas-Liquid Reactions," McGraw-Hill, New York, 1970.

Downing, A. L. et al.: "The Performance of Mechanical Aerators," Jour., Inst. Sewage Purif., Pt. 3, 231 (1960).

Drews, R. J. et al.: "The Orbal Extended Aeration Activated Sludge Plant," Jour. Water Poll. Control Fed., 44, (1972).

Eckenfelder, W. W. et al.: "Effects of Various Organic Substances on Oxygen Absorption Efficiency," Sewage and Ind. Wastes, 28, 1357 (1956).

von der Emde, W.: "Die Technik der Belueftung in Belebtschlammanlagen," paper presented at the EAWAG Conference, Zurich (Switzerland), April, 1964.

Ettlich, W. F.: "A Comparison of Oxidation Ditch Plants to Competing Processes for Secondary and Advanced Treatment of Municipal Wastes," EPA-600/2-78-051 (1978).

Fischerstroem, N. C. H.: "Low Pressure Aeration of Water and Sewage," Proc. Amer. Soc. Civ. Eng., Jour. San. Eng. Div., 86, SA5, 2607 (1960).

Flanagan, M. J. et al.: "Automatic Dissolved Oxygen Control," paper presented at the ASCE Environmental Engineering Division Conference on Research, Development and Design, Seattle, Washington, 1976.

Flanagan, M. J. and Bracken, B. D.: "Design Procedures for Dissolved Oxygen Control of Activated Sludge Processes," EPA-600/2-77-032, Cincinnati, Ohio, June 1977.

Gameson, A. L. H., and Robertson, K. G.: "The Solubility of Oxygen in Pure Water and Sea Water," Jour. Appl. Chem. (Brit.), 5, 502 (1955).

Ganczarczyk, J.: "Some Features of Low-Pressure Aeration," paper presented at the 2nd Intern. Conf. on Water Poll. Research, Tokyo, Japan, 1964.

Ganczarczyk, J., and Suschka, J.: "Pilot-Plant Studies on Coarse Bubble Aeration," Intern. Jour. Air Water Poll., 6, 319 (1962).

Gaudy, A. F., et al.: "Biological Treatment of Volatile Waste Components," Jour. Water Poll, Control Fed., 35, 75 (1963).

Gibbon, D. L. (editor): "Aeration of Activated Sludge in Sewage Treatment," Pergamon Press, New York, 1974.

Hermanowicz, S. W., and Ganczarczyk, J.: "Some Hydrodynamic Characteristics of Three-Phase Fluidized Beds," paper presented at the 16th Canadian Symposium on Water Pollution Research, Toronto, Ontario, 1981.

Hsu, K. H. et al.: "Motionless Mixers Improve Oxygen Transfer Efficiencies," Water & Sewage Wks, 122, 8, 76 (1975).

Hurd, C. H.: "Design Features of the Indianopolis Activated Sludge Plant," Eng. News-Red., Aug. 16, (1923).

Husmann, W.: "Belueftungsbecken fuer Belebtschlammklaeranlagen," German Patent No. 545.897 of February 18, 1932.

Ippen, A. T. et al.: "The determination of Oxygen Absorption in Aeration Processes," Massachusetts Institute of Technology Hydrodynamics Lab., Technical Report No. 7, May (1952).

Jennekens, H.: "Water-Jet Technique Combines Aeration and Mixing," Water & Sewage Wks, 126, 3, 71 (1979).

Jeris, J. S. et al.: "Biological Fluidized Bed Treatment for BOD and Nitrogen Removal," Jour. Water Poll. Control Fed., 49, 816 (1977).

Kaetin, J. R.: "Einrichtung zum Umwaelzen und Beluefften von Fluessigkeit", Swiss Patent No. 2442 of February 27, 1964.

Kessener, H., and Ribbius, F. J.: "Comparison of Aeration Systems for the Activated Sludge Process," Sewage Works Jour. 6, 423 (1934).

Knopp, E. et al.: "Versuche mit verschiedenen Belüftungssystemen. Teil 1. Untersuchungen an Oberflachenbelüftern und Kombinierten Systemen in Reinwasser," Vulkan Verlag, Essen 1964.

Leary, R. D., et al.: "Full-Scale Oxygen Transfer Studies of Seven Diffuser Systems," Jour. Water Poll. Control Fed., 41, 459 (1969).

Lewis, W. K., and Whitman, W. G.: "Principles of Gas Absorption," Ind. and Eng. Chem, 16, Dec. (1924).

McKinney, R. E., and Conway, R. A.: "Chemical Oxygen in Biological Waste Treatment," Sewage and Ind. Wastes, 29, 1097 (1957).

McWhirter, J. R. (editor): "The Use of High-Purity Oxygen in the Activated Sludge Process," Vol. I & II, CRC Press, 1978.

Micka, T. et al.: "Reaeration Experiments with Micro-organisms," Proc. Am. Soc. Civ. Eng., Jour. Env. Eng. Div., $\underline{99}$, EE6, 971 (1973).

Miyaji, Y., and Kato, K.: "Biological Treatment of Industrial Waste Water by Using Nitrate as an Oxygen Source," Water Research (Brit.), $\underline{9}$, 95 (1975).

Muskat, J.: "Ueber die Weiterentwicklung der Kessener Buerste zu modernen Rotoren hoeher Leistung," Muencher Beitr. (Ger.), $\underline{5}$, 164 (1958).

Muskat, J.: "Operating Experiences with Mammoth Rotors in Circulation Tanks," Prog. Wat. Tech. (Brit.), $\underline{11}$, No. 3, 111 (1979).

Neal, L. A., and Tsivoglou, E. C.: "Tracer Measurement of Aeration Performance," Jour. Water Poll. Control Fed., $\underline{46}$, 2, 247 (1974).

Nelson, F. G.: "Spiral Contact Aeration in Biological Treatment," Sewage Ind. Wastes, $\underline{30}$, 907 (1958).

Owen, W. F., and Slechta, A. F.: "Organic Removal or Nitrification with a Combined Fixed-Suspended Growth Biological Treatment System," paper presented at the 48th Annual Conference of the Water Pollution Control Federation, Miami Beach, Florida, 1975.

Pasveer, A.: "Research on Activated Sludge," Sewage and Ind. Wastes $\underline{25}$, 12 (1956) and $\underline{28}$, 28 (1958).

Paulson, W. L., and Lively, L. D.: "Oxidation Ditch Treatment of Meatpacking Wastes," EPA-600/2-79-030 (1979).

Pearlman, S. R., and Fullerton, D. G.: "Full-Scale Demonstration of Open Tank Oxygen Activated Sludge Treatment," EPA-600/2-79-012, May 1979.

Roesler, J. F.: "Plant Performance Improvement Using Dissolved Oxygen Control," Proc. Amer. Soc. Civ. Engrs., Jour. Envir. Eng. Div., $\underline{100}$, EE5, 1069 (1974).

Rose, W. L., and Gorringe, R. E.: "Deep-tank Extended Aeration of Refinery Wastes," Jour. Water Poll. Control Fed., $\underline{46}$, 393 (1974).

Schellmann, E.: "Bayer Tower Biology - An Efficient, Low Noise and Odour Free Process for Biological Waste Water Treatment," paper presented at the Xth IAWPR Conference, Toronto, June 1980.

Smart, J.: "Oxygen Transfer Testing Procedures in Ontario," paper presented at "Workshop 80", University of Toronto, Toronto, April 1980.

Speece, R. E.: Personal information, 1979.

Speece, R. E. et al.: "Pilot Performance of Deep U-Tubes," paper presented at the Xth IAWPR Conference, Toronto, June 1980.

Stalmann, V.: "Die BSK-Turbine-ein neues Hochleistung-Belueftungssystem der Abwasser Technik," Gas-und Wasserfach (Germany), 106, 613 (1965).

Stalzer, W., and von der Emde, W.: "Tanks with Turbulent Flow Generated by Mammoth Rotors," Water Research (Brit.), 6, 417 (1972).

Stetzer, R. H.: "The MAROX System - Pure Oxygen Wastewater Treatment in Open Tanks," paper presented at the 48th Annual Conference of Water Pollution Federation, Miami Beach, Florida, October 1975.

Steward, M. J., and Lidkea, T. R.: "Vertical Static Tube Aerators: Evaluating Their Performance," Water & Sewage Wks., Reference Number 80 (1976).

Tischler, L. F., et al.: "Evaluation of the Potential for Air Stripping of Hydrocarbons during Activated Sludge Wastewater Treatment," paper presented at the 51st Annual Conference of Water Pollution Control Fed., Anaheim, California, 1978.

Toerber, E. D., and Mandt, M. G.: "Greater Oxygen Transfer with Jet Aeration System," Water & Sewage Works, 126, 1, 71 (1979).

Tsao, G. T.: "Simultaneous Gas-Liquid Interfacial Oxygen Absorption and Biochemical Oxidation," Biotech. and Bioeng. 10, 765-785 (1968).

Water Pollution Control Federation: "Aeration in Wastewater Treatment," Manual of Practice No. 5, Washington, D.C., 1971.

Weston, R. F.: "Studies in Entrainment Aeration," Jour. Water Poll. Control Fed., 34, 342 (1962).

Wilford, J., and Conlon, T. P.: "Contact Aeration Sewage Treatment Plants in New Jersey," Sewage and Ind. Wastes, 29, 845 (1957).

Wong-Chong, G. M., et al.: "Comparison of the Conventional Cage Rotor and Jet-Aero-Mix in Oxidation Ditch Operation," Water Research (Brit.), 8, 761 (1974).

Zeper, J., and de Man, A.: "New Development in the Design of Activated Sludge Tanks with Low BOD Loadings," paper presented at the 5th International Conference on Water Pollution, San Francisco, California, 1970.

Activated Sludge Separation and Thickening

6.1 SEPARATION PURPOSE AND SEPARATION METHODS

An integral part and the most important limiting factor of activated sludge process treatment is a separation of biological solids from the treated wastewater. For the purpose of sludge recycling and effluent clarification, sedimentation (see 6.2) is being used mostly, but the application of flotation (see 6.3) is also gaining some popularity. Both gravity sedimentation and flotation are also used for sludge thickening. Centrifuging of mixed liquor is not sufficient as a single separation method, but can be used effectively for sludge thickening (see 6.4). A special type of centrifugal screen concentrator (see 6.5) may be used also for sludge thickening for the purpose of recycling. For removal of residual suspended solids from the clarification effluent, micro-screening or filtration (see 6.6) are used.

6.2 GRAVITY SEPARATION AND THICKENING

Gravity separation of activated sludge is achieved by sedimentation of sludge flocs that have a higher specific gravity than the clarified liquid. This operation calls for relatively quiescent conditions which are achieved for this purpose in settling basins. These are frequently called secondary clarifiers as, historically, this sedimentation operation in many sewage treatment plants was preceded by sedimentation of raw sewage suspended solids in primary clarifiers. Activated sludge flocs are flocculating during the operation of sedimentation, and this phenomenon is influenced among other things, by the level of suspended solids in mixed liquor. Therefore, both the rate of particles growth and the initial concentration of suspended solids affect the performance of the operation which is controlled by overflow rate, detention time, solids loading, the physical features of the tank, and temperature.

Gravity separation of activated sludge is associated with some thickening of the separated solids. However, the conditions for the best performance of one of these operations often have a hindering effect on the other. Gravity sludge thickening performance in secondary clarifiers used to be estimated empirically on the basis of sludge volume index tests (see Chapter III), but this approach had serious limitations as it did not refer to the specific hydraulics of the clarifiers. An alternative approach of the "solids-flux" method is described later.

Performance of the secondary clarifiers can be expressed in terms of the efficiency in removing suspended solids and in terms of the sludge compaction characteristics. They should be sized adequately for both clarification and thickening functions (Dick, 1970). Many malfunctions of activated sludge systems appear to be due rather to settling or thickening problems than to biological transformations. The inability of gravity separation to produce an effluent virtually free of biological growth may be due to several factors: an excessive sludge volume index (see Chapter III, p. 3.5), denitrification of wastewater in secondary clarifiers (see Chapter X, p. 10.5), hydraulic overloading, excessive solids loading, poor design of units, etc. Addition of some polymers, in doses from 0.2 to 5 mg/L, to mixed liquor prior to secondary clarification may effectively decrease the residual level of suspended solids in the final effluent, increase the sludge settling rate and increase the underflow sludge concentration. However, application of such chemical additions should be justified technologically and economically. Poduska et al. (1979) presented a successful pilot-plant scale and full-scale case study of such an approach.

SETTLING AND THICKENING PROPERTIES
OF ACTIVATED SLUDGE

Sedimentation of activated sludge mixed liquor suspended solids is classified as so-called "hindered settling" in which the settling rate is a function of the suspended solids concentration, and in which both detention time and solids loading govern the operation. The settling and thickening characteristics of activated sludge vary substantially with the modifications of the process, type of wastewater, and the mixed liquor concentration. Generally, settling velocities decrease as the mixed liquor suspended solids concentration increases. Dick and Young (1972) described this relationship using the following empirical equation:

$$v_i = aC^{-b} \tag{6-1}$$

where: v_i = initial settling velocity, "a" and "b" = empirical constants, and "C" = mixed liquor suspended solids concentration. A somewhat different, empirical equation for initial settling velocity was also proposed by Schaffner and Pipes (1978). The constants for these equations can be

obtained from laboratory or field studies. A range of observed initial
settling velocity values (without addition of chemicals for phosphorus precipitation) is presented in Fig. 6-1. It must however, be stressed that use
of this information for selection of the clarifier overflow rate is theoretically incorrect (Dick, 1976).

Application of the hydraulic theory of limiting-flux clarifiers' design
is based on a concept established by Kynch (1952) and applied to activated
sludge mixed liquor first by Eckenfelder and Melbinger (1957) and then, in
a more practical way, by Dick (1970) and Dick and Young (1972). According to this theory, for a particular type of sludge, there is an operation-
limiting solids flux that can be applied in a given clarifier. This flux can
be determined by graphical or mathematical methods (Dick and Young,
1972).

Process loading parameters influence sludge clarification and thickening characteristics. Information on the subject was reported by Ford and
Eckenfelder (1967), Foster (1968), Bisogni and Lawrence (1970), Chao and
Keinath (1979), and others. However, many aspects of the respective
phenomena still are not completely clear.

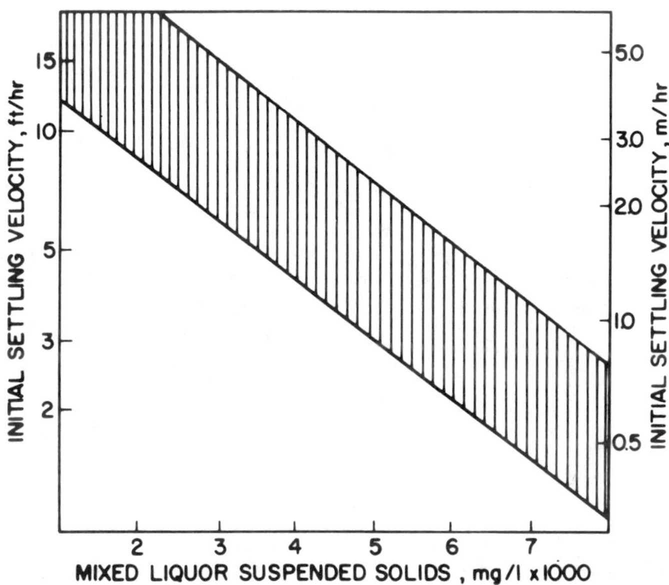

Figure 6-1. Influence of mixed liquor concentration on activated sludge
sedimentation rate.

HYDRAULICS OF SETTLING AND THICKENING

In most clarifiers and thickeners, the settling and thickening operation is
affected by hydraulic conditions existing in the respective units. This includes flow variations, design of the inlet and the outlets (to provide for
horizontal and vertical distribution to minimize short circuiting and turbulence), as well as the formation of density currents, sludge blankets, etc.
To eliminate wind and thermal effects on the operation of secondary clarifiers, various types of covers have been used (see Chapter IX, p. 9.5).
An unusual type of cover for a circular clarifier may be the placing of
small plastic ping pong balls on its surface.

SECONDARY CLARIFIERS

The sedimentation units in activated sludge treatment can be either circular
(radial or square radial) or rectangular, and their volume can be divided
into inlet zone, clarification zone, outlet zone and sludge zone. Separation
of activated sludge flocs takes place in the clarification zone, and accumulation and thickening of sludge in the sludge zone. Circular clarifiers can be
centre feed or peripheral feed. Figure 6-2 presents a common type of
centre-feed circular clarifier.

Secondary clarifiers for the activated sludge process are usually designed on the basis of the suggested maximum overflow rates, minimum
retention times and maximum sludge loadings (Table 6-1). These conditions, however, often do not guarantee the expected performance of the
units. It appears that there still is insufficient knowledge on the subject.
Table 6-2 presents the recommended solids loading to achieve a residual
suspended solids level of 30 mg/L at different sludge volume index values
(Pflanz, 1969).

A mathematical model for secondary clarifiers was developed by the
U.S. Environmental Protection Agency (1972). The performance of circular clarifiers was studied by Villemonte and Rohlich (1963), Munch and
Fitz-Patrick (1978), and others. Tube settlers were the subject of investigations by Culp et al. (1969), Slechta and Conley (1971), and Mendis and
Benedek (1980). Design and application of lamella clarification were presented by Grimes and Nyer (1978). More than one-stage settling was
studied by Lee et al. (1976 and 1977), and multi-storey horizontal settling
tanks were described by Fischerstorm et al. (1967) and Matsunaga (1980).

Integral clarifiers for activated sludge systems represent a special
type of sedimentation unit (see Chapter V, p. 5.7). The most advanced
modification of such a design is based on patent claims by Love (1974).
Fig. 6-3 presents an internal view of an integral clarifier (scraping installation), and Fig. 6-4 and 6-5 show a treatment plant of this type during
construction and under operation. Clarifiers with a separate sludge thickening facilities were also described by Love (1981).

Figure 6-2. Secondary circular clarifiers in the Paris-Acheres Sewage
Treatment Plant (France)—photo by author.

SLUDGE REMOVAL FROM SECONDARY CLARIFIERS

Effective sludge removal from secondary clarifiers is an important operation as, under certain conditions (nitrifying sludge), clarifiers should be as
free of sludge as possible. Sludge can be removed from secondary
clarifiers either by scraping (Fig. 6-6) or by suction (Fig. 6-7). Scrapers
are used both in rectangular clarifiers (chain scrapers or travelling-bridge
scrapers) and in circular clarifiers (raking mechanism with a number of
blades per collector arm). The latter operation is considered to be less
suitable as sludge has to be moved a few times from one blade to another
in the raking mechanism. It is thought that the scrapers operate by some
resuspending of sludge and cause it to flow as a thixotropic liquid to the removal point. Devices for sludge removal by suction can be used in rectangular clarifiers equipped with a travelling bridge; in circular clarifiers,
rotating tapered headers or multiple-pipe drawoffs are used.

It appears that suction sludge removal is more advantageous than
scraping removal as the sludge can be removed faster this way, without
too much agitation.

Application of surface skimmers is recommended for most secondary
clarifiers, operating under specific conditions. It prevents deterioration

TABLE 6-1
Recommended Loadings for Activated Sludge Secondary
Clarifiers
(after Task Committee on Final Clarification, 1979)

Loadings	Average	Peak
Hydraulic Loadings		
in gpd per sq. ft.	400–800	1,000–1,200
in cu. m per sq. m per day	16–32	40–48
Solids Loadings		
in lbs per sq. ft per day	20–30	< 50
in kg per sq. m per day	98–146	< 244

of the secondary effluent quality, caused by suspended solids floating to
the surface of the clarifiers when a limited denitrification of nitrified
wastewater occurs due to development of anoxic conditions in the clarifiers. In the extended aeration process, there is often a tendency for some

TABLE 6-2
Recommended Solids Loadings to Achieve the Residual
Suspended Solids Level of 30mg/L at Different Sludge
Volume Index (SVI) Values
(after Pflanz, 1969)

SVI, mL/g	Solids Loadings per hr	
	lbs per sq. ft	kg per sq. m
100	0.7	3.5
200	0.2–0.3	1.1–1.3
300	0.1–0.2	0.8–1.1

Figure 6-3. Internal view of an integral clarifier (photo courtesy of Greey Lightnin).

non-biodegradable material to accumulate on the surface of the clarifiers. This material should be removed by surface skimming.

SEPARATE GRAVITY THICKENING

Thickening of activated sludge occurs as a part of the separation of the activated sludge from the mixed liquor in the secondary clarifiers; or it is practised independently as an additional operation, usually for mixed sludges (primary and secondary). Separate gravity thickening of activated sludge only is not a very effective operation. At a relatively low thickener solids surface loading of 4-6 lb/day/sq. ft. (20-30 kg/m^2/day), thickening to only

Figure 6-4. Integral clarifier activated sludge plant under construction in Barry's Bay, Ontario (photo courtesy of Greey Lightnin).

2.5-3% of the solids may be achieved. Sometimes the thickening results are even lower, with a maximal concentration to only 1.2-2% solids. However, oxygen-activated sludge or some sludges from the extended aeration modification of the process, may produce sludges concentrated to 4% solids. A formula was offered for calculation of the underflow concentration of activated sludge gravity thickener. The expected concentration in % solids is 1.75 + 1/SVI, where SVI = sludge volume index in mL/g. However, the reliability of this formula is questionable. George and Keinath (1978) developed a mathematical model simulating the dynamic response of a continuous sludge thickener to various external forces.

In gravity thickening of excess activated sludge, the addition of polyelectrolytes is usually considered. Such thickening may be practised as the first step towards sludge stabilization by aerobic digestion. However, when anaerobic stabilization of sludge is planned, it is often better to thicken by gravity the combined sludge (primary sludge and excess of activated sludge), at a loading in the range of 6-10 lb/day/sq. ft. (30-50 kg/m^2/day), usually without the addition of polyelectrolytes. A concentration in the range of 3-5% solids or more, can be achieved for the resulting sludge.

For separate gravity thickening, circular tanks are normally used with a side water depth of 10 to 12 ft (3-3.7 m) and a relatively steep floor slope. Mechanical stirrers are required for thickening of heavy sludges but, for excess activated sludge only or for primary sludge and excess activated sludge, they may not be necessary. In Sudbury, Ontario, excess activated sludge from a plant operating without primary clarification, is thickened in an up-flow clarifier with the addition of polyelectrolytes. with the addition of polyelectrolytes, is thickened in an up-flow clarifier. The flow through the thickened sludge blanket which is formed, produces sludge of 4% concentration. This technical modification of the thickening operation may develop into an attractive alternative to the conventional thickeners.

6.3 SEPARATION AND THICKENING BY FLOTATION

Flotation may be used for separation of activated sludge from mixed liquor or for thickening of sludge separated from mixed liquor by gravity. The first application has not gained much popularity except for use in the deep shaft system (see Chapter IV), but activated sludge thickening by flotation has become quite a common operation. This operation is based on inducing

Figure 6-5. Integral clarifier activated sludge plant under operation in Barry's Bay, Ontario (photo courtesy of Greey Lightnin).

Figure 6-6. Chain scraper in a rectangular clarifier.

fine air bubbles in mixed liquor or settled sludge. These bubbles attach
themselves to the suspended solids and thus cause them to rise to the surface. The floated particles are then collected by a skimming operation.
Fine air bubbles in the mixed liquor can be formed by compressed air diffusion, depressurizing the liquids saturated with air under pressure, and
application of a vacuum to liquids saturated with air under atmospheric
pressure. The second method, called dissolved-air flotation (DAF) is the
most common. It has two distinct modifications: the entire flow of mixed
liquor can be pressurized with air (used only for small units) or a portion

of the effluent can be recycled, pressurized with air, and mixed with the
unpressurized main flow (Fig. 6-9).

The factors affecting the performance of activated sludge flotation are:
type of sludge and equipment used, sludge loading rate, air to solids ratio
and flow detention time (Eckenfelder et al., 1958; Katz and Geinapolos,
1964; Mulbarger and Huffman, 1970):

Sludge Loading Rate, usually expressed in lb (kg) of solids per sq. ft.
(sq. m) of flotation units per day. This factor is commonly used as
one of the main design parameters for the operation.

Air/Solids Ratio, expressed in lb (kg) per lb (kg), is another operation
design parameter used often in combination with the sludge loading
rate. This ratio considers several individual variables of flotation,
including the effect of temperature on the solubility of air, recycle
volume and pressure, and feed solids concentration and volume.

Float Detention Time, defined as the time required for the surface
sludge collectors to traverse the length of the tank. An increase in the

Figure 6-7. Scrapers in a circular clarifier.

Figure 6-8. Sludge removal by suction in a circular clarifier.

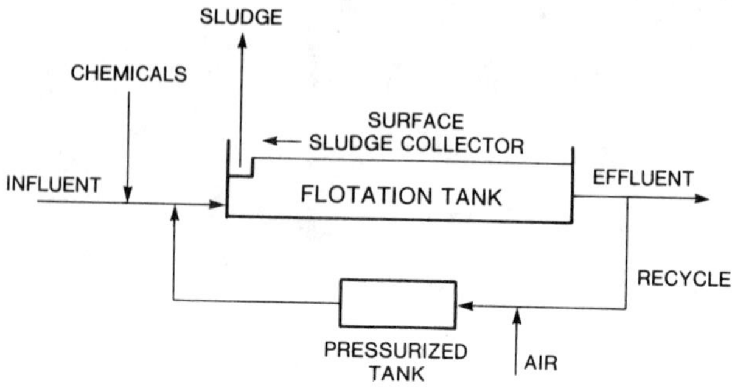

Figure 6-9. Activated sludge thickening by dissolved air flotation.

float detention time causes an increase in the sludge concentration,
but too long a detention time can cause a decrease in the sludge float-
ability, due to the oxygen uptake by the sludge.

The dissolved-air flotation system is very flexible, as the recycle and
the applied pressure for air solubilization can be controlled within a broad
range. Usually, the recycle is pressurized to 40-80 psi (2.8-5.6 kg per
sq. cm) with the excess of air, and the bubbles formed after mixing and
depressurizing have diameters of 80μ or less. The majority of the flotation
thickeners operate with experimentally selected additions of polyelectrolytes. These are similar to chemicals used to improve gravity sedimentation or thickening. The addition of the chemicals may affect both the concentration of the float and the supernatant residual suspended solids.
Normally, it is necessary to at least carry out laboratory experiments to
develop fully the design criteria for activated sludge flotation, although
often the full-scale installations perform better than laboratory units. The
laboratory equipment for evaluation of activated sludge flotation was developed by Wood and Dick (1973) and modified by Gehr and Henry (1976). The
mode of operation of the activated sludge treatment and specifically sludge
age affect the sludge flotation properties (Gulas et al., 1978).
Typical design criteria for flotation thickening of excess activated
sludge without the use of chemicals call for surface loadings of 8 to 15 lb
per day per sq. ft. (39-73 kg per day per sq. m), with an expected float
concentration between 3 and 6 percent. For mixed primary and activated
sludges, the respective loadings are in the range from 20 to 30 lb per day
per sq. ft. (98-146 kg per day per sq. m) and the concentrations in the
range from 5 to 10 percent. In the plant studies by Burfitt (1975), a 4 percent float was produced from 5,000 mg/L excess activated sludge, with
polyelectrolyte addition at loadings of 2 lb per hr per sq. ft. (9.75 kg per
hr per sq. m) or less. Detailed design procedures for dissolved-air
flotation systems were described by Bratby and Marais (1976 & 1977).

6.4 CENTRIFUGAL THICKENING

Centrifuges are not used for the separation of activated sludge from mixed
liquor, but their application for thickening of activated sludge is common.
The mechanism of centrifugal thickening of waste activated sludge alone or
with primary sludge, is analogous to thickening by gravity. However, the
applied centrifugal force is usually 500 to 3,000 times greater than the
force of gravity. The basic parameters of the operation are the type of
centrifuge, its rotational speed, and the sludge retention time. The effects
of thickening are measured as concentration of the cake solids and solids
capture. For most of the sludges, the effective results cannot be achieved
without the addition of polymers.
Three types of centrifuges are used for sludge thickening: imperforate
basket centrifuge, disk-nozzle centrifuge, and solid-bowl conveyor centri-

Figure 6-10. Laboratory equipment for evaluation of activated sludge
flotation.

fuge. As all these centrifuges are sensitive to the grit contents of the sludge which is being thickened, appropriate sludge pretreatment is normally required. Excess activated sludge received from the process modifications operating after primary clarification of wastewater, does not contain any meaningful quantities of grit and its degritting may not be necessary. From other sludges, an effective grit separation can be achieved through the use of hydrocyclones capable of removing grit of 150 to 200 mesh particle size (65-75μ m) and larger.

Various activated sludges respond in a very individual manner to the centrifugal thickening, and it is advisable to design the respective systems on the basis of thickening experiments made on the specific sludges.

Kennedy et al. (1968) studied the effect of returning centrate from a centrifuge used to concentrate waste activated sludge. They found that the centrate contained broken floc particles and bacterial filaments. Most of the solids in the centrate were removed by sedimentation, but the bacterial filaments carried back into the aeration tank caused higher populations of filamentous bacteria in the activated sludge.

6.5 CENTRIFUGAL SCREEN CONCENTRATOR

The centrifugal screen concentrator for activated sludge mixed liquor (Fig. 6-11) is a combination of a microstrainer and a centrifuge (Fernbach and Tchobanoglous, 1975). In such a unit, mixed liquor is pumped into a rotating screen cage. Most of the liquid and some of the solids pass through the rotating screen, while the majority of the solids are separated. The screens have a typical mesh size from 165 to 400 (105 to 37 micron) and rotate at speeds depending on the size of the unit and varying from 60 to 350 rpm. A backwash spray is directed to the outside of the screen. The application of centrifugal screen concentrators can effectively change the conventional activated sludge flowsheet. By returning the screen concentrate to aeration tanks and directing screen effluent to the secondary settling tanks (Fig. 6-12) it is possible to maintain higher solids concentrations in aeration tanks and/or to improve the performance of the settling tanks by decreasing the respective solids loadings. Moreover, wasting of sludge can be accomplished as removal of a portion of the screen concentrate. The performance of centrifugal screen concentration can be characterized on the basis of: (1) the degree of solids concentration achieved (defined as the ratio of the suspended solids in the screen concentrate to the suspended solids in the screen effluent), (2) flow split (defined as the ratio of the flow passing through the screen to the influent flow), and (3) solids capture (defined as the percentage of the influent solids mass that is discharged in the screen concentrate stream). Experiments performed at four California plants showed that it was possible to achieve in this way a degree of solids concentration in the range from 2.08 to 7.44 times (e.g., to 20,000 mg/L), a flow split of 80 percent, and a solids capture in the range from 54 to 79 percent.

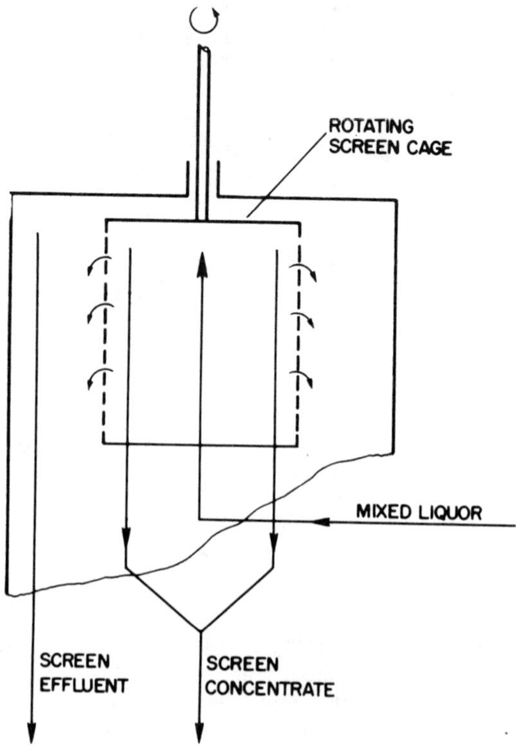

Figure 6-11. Centrifugal screen concentrator (idealized).

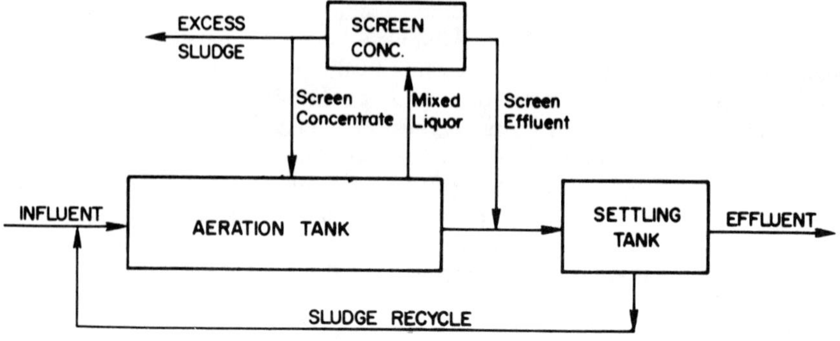

Figure 6-12. Application of centrifugal screen concentrator in activated
sludge process (after Fernbach and Tchobanoglous, 1975).

6.6 EFFLUENT MICROSTRAINING AND FILTRATION

Effluents from the secondary clarifiers of properly designed and operated
activated sludge treatment plants often have a suspended solids content of
about 15 mg/L. A requirement to keep the level of suspended solids in the
discharge to the receiving surface waters at 5 or 6 mg/L, for example,
calls for additional treatment by microstraining or filtration.
<u>Microstrainers</u> consist of a rotating drum with a peripheral screen. In-
fluent wastewater enters the drum internally and passes radially outward
through the screen, with deposition of solids on the inner surface of the
drum screen. At the top of the drum, pressure jets remove the deposited
solids. The backwash water is then collected and returned to the head of
the plant. The screens employed in microstrainers have extremely small
openings and are made from a variety of metal or plastic fibers (e.g.,
23-micron or 35-micron fabrics). One of the advantages of using a microstrainer is its low head requirement. Head loss through the microstraining unit, including inlet and outlet structures is about 12 to 18 inches
(30.5-46 cm). Across the screen, a 6-inch (15 cm) limit is usually imposed at peak flows. Head loss build-up is reduced by increasing the rate
of drum rotation and by increasing the pressure and flow of the backwash-
ing jets. Characteristic case studies of the application of microstraining
to the activated sludge plant effluents were presented by Lyman et al.
(1969), Daiper (1973) and others. Ewing (1976) formulated design criteria
for the respective units.
<u>Filtration</u> of activated sludge plant effluents is practised in different types
of facilities, operated in different modes, and is often supported by the
addition of chemicals. Generally, the filtration effects are a function of
media type, size, and depth, filtration rate, filter influent quality, and
the addition or presence of specific polyelectrolytes. Some biological
activity in these tertiary filters is often possible, and may affect the
treatment results by also removing some soluble BOD from the wastewater.

Tchobanoglous (1970) studied the application of granular media filters,
for secondary effluents and reported that, at a depth from 18 to 30 in.
(46-76 cm), media grain size had a significant effect on particles removal
efficiency for single medium sand and anthracite. Influent suspended solids
studied varied from approximately 14 to 24 mg/L, and small removals
occurred below depths of 16 to 20 in. (41-51 cm). Other studies by Hunsinger and Rupke (1975) showed that, at hydraulic loads in the range '
2.4-9.6 gpm per sq. ft. (98-391 L per min. per sq. m) with media consisting of 36 in. (91 cm) of anthracite of 0.9 mm effective size (uniformity
coefficient = 1.65) and 12 in. (30 cm) of sand of 0.5 mm effective size
(uniformity coefficient = 1.5), a consistently high quality effluent was produced with a residual suspended solids concentration below 5 mg/L. About
3.2 percent of the effluent was used for backwash. Particle size analysis
indicated that filtration was more efficient in removing particles greater
than 14.4 microns. To effectively remove smaller particles, a coagulation
of the filter feed would be required. It was also observed that higher fil-

tration rates caused a shearing of suspended solids into smaller particles. In these experiments, backwash was accomplished automatically when the filter reached 60 in. (152 cm) of head loss above initial dynamic head loss. The backwash sequence consisted of 5 min. draindown time, 5 min. of air scour, and a 1 min. refill period, followed by 5 min. of backwash at 15.6 gpm per sq. ft. (635 L per min. per sq. m).

Swanson and FitzPatrick (1978) and Fitzpatrick and Swanson (1980) found, in their studies on tertiary filters, that simple correlations between filter solids removal efficiency and filter operating parameters, such as flow and solids loading, were generally poor. Differences in the characteristics of secondary effluent suspended solids rather than media grain size and depth were most likely the major reasons for variations in clarification efficiency between particular applications of filters. The most important design consideration for such filters proved to be the ability to handle shock loads caused by activated sludge process upsets.

A less common application of the filtration concept to upgrade activated sludge process effluents is filtration through polyurethane foam (Ramagoli, 1978). The advantages of this filtration media over the more conventional materials are lack of head loss buildup for streams with high (more than 100 mg/L) influent suspended solids levels, higher solids loading capacity and longer run times.

REFERENCES

Bisogni, J. J., and Lawrence, A. W.: "Relationship Between Biological Solids Retention Time and Settling Characteristics of Activated Sludge," Water Research (Brit.), 5, 753 (1970).

Bratby, J., and Marais, G. R.: "A Guide for the Design of Dissolved-Air (Pressure) Flotation Systems for Activated Sludge Processes," Water SA, 2, 87, No. 2 (1976).

Bratby, J., and Marais, G. V. R.: "Dissolved-Air Flotation in Activated Sludge," Prog. Wat. Technol., 9, 311 (1977).

Burfitt, M. L.: "The Performance of a Full-Scale Sludge Flotation Plant," Water Pollution Control (Great Britain), 74, 474 (1975).

Chao, A. C., and Keinath, T. M.: "Influence of Process Loading Intensity on Sludge Clarification and Thickening Characteristics," Water Research (Brit.), 13, 1213 (1979).

Culp, G. L. et al.: "Tube Clarification Process - Operating Experiences," Jour. San. Eng. Div., Amer. Soc. Civil Engrs., 95, SA5, 829 (1969).

Daiper, E. W. J.: "Tertiary Treatment by Microscreening - Case Histories," Water and Sewage Wks., 120, 8, 42 (1973).

Dick, R. I.: "Role of Activated Sludge Final Settling Tanks," Jour. San. Eng. Dir., Proc. Amer. Soc. Civ. Eng., 96, SA2, 423 (1970).

Dick, R. I., and Young, K. W.: "Analysis of Thickening Performance of Final Settling Tanks," Proc. 27th Purdue Ind. Waste Conference, Lafayette, Indiana, 1972.

Dick, R. I.: "Folklore in the Design of Final Settling Tanks," Jour. Water Poll. Control Fed., 48, 633 (1976).

Eckenfelder, W. W., Jr., et al.: "Studies on Dissolved Air Flotation of Biological Sludges" in "Biological Treatment of Sewage and Industrial Wastes," Vol. II, "Anaerobic Digestion and Solids - Liquid Separation," Reinhold Publishing Company, New York, 1958.

Eckenfelder, W. W., Jr., and Melbinger, N.: "Settling and Compaction Characteristics of Biological Sludges," Sewage and Ind. Wastes, 29, 1114 (1957).

Environmental Protection Agency: "A Mathematical Model of a Final Clarifier," Project No. 17090F5W (1972).

Environmental Protection Agency: "Process Design Manual for Suspended Solids Removal," EPA 625/1-75-033, January 1975.

Ewing, L.: "Design Criteria for Microscreening Biological Wastewater Plant Effluents," Water and Sewage Wks., Reference Number, 94 (1976).

Fernbach, E., and Tchobanoglous, G.: "Centrifugal Screen Concentrator for Activated Sludge," Water & Sewage Wks., 122, 1, 64 and 122, 2, 40 (1975).

Fischerstörm, C. N. H. et al.: "Settling of Activated Sludge in Horizontal Tanks," Proc. Amer. Soc. Civ. Engrs., Jour. Sanit. Eng. Div., 93, SA3, 73 (1967).

FitzPatrick, J. A., and Swanson, C. L.: "Evaluation of Full-Scale Tertiary Wastewater Filters," EPA-600/2-80-005, Cincinnati, Ohio, May 1980.

Ford, D. L., and Eckenfelder, W. W., Jr.: "Effect of Process Variables on Sludge Floc Formation and Settling Characteristics," Jour. Water Poll. Control Fed., 39, 1850 (1967).

Forster, C. F.: "The Surface of Activated Sludge Particles in Relation to Their Settling Characteristics," Water Research (Brit.), 2, 767 (1968).

Gehr, R., and Henry, J. G.: "Laboratory Techniques for Evaluating Batch Flotation of Activated Sludge," paper presented at the XI Canadian Symposium on Water Pollution Research, Burlington, Ontario, February 1976.

George, D. B., and Keinath, T. M.: "Dynamics of Continuous Thickening," Jour. Water Poll. Control Fed., 50, 2560 (1978).

Grimes, C. B., and Nyer, E. K.: "Lamella Clarification: Design-Applications," Proc. 33rd Purdue Ind. Waste Conference, Lafayette, Indiana, 1978.

Gulas, V., et al.: "Factors Affecting Design of Dissolved Air Flotation Systems," Jour. Water Poll. Control Fed., 50, 1853 (1978).

Henry, J. G., and Gehr, R.: "Dissolved Air Flotation for Primary and Secondary Clarification," Report to Central Mortgage and Housing Corporation, Toronto, 1977.

Hunsinger, R. B., and Rupke, J. W. G.: "A Preliminary Report on Pilot-Plant Filtration of Secondary Effluents," paper presented at the Seminar: "High Quality Effluents," Toronto, 1975.

Katz, W. J., and Geinapolos, A.: "Sludge Thickening by Dissolved-Air Flotation," Jour. Water Poll. Control Fed., 36, 407 (1964).

Kennedy, T. J., et al.: "Effect of Centrate from Solid Bowl Centrifuge on Activated Sludge Process," Proc. 22nd Ind. Waste Conf., Purdue Univ., Ext. Ser. 129, 145 (1968).

Kynch, G. J.: "A Theory of Sedimentation," Trans. Faraday Society, 48, 166 (1952).

Lee, C. R., et al.: "Two-Stage Settling Improves Sludge Removal Efficiency," Water & Sewage Works, 41, May, 1977.

Lee, C. R., et al.: "Optimization on Multistage Secondary Clarifier," Jour. Water Poll. Control Fed., 48, 2578 (1979).

Love, L. S.: "Treatment Apparatus," U.S. Patent 3,778,477 of Jan. 29, 1974.

Love, L. S.: "Liquid Treatment Apparatus," U.S. Patent 4,272,369 of June 9, 1981.

Lyman, B. T., Etteet, G. and McAloon, T.: "Tertiary Treatment at Metro Chicago by Means of Rapid Sand Filtration and Microstrainers," Jour. Water Poll. Control Fed., 41, 247 (1969).

Matsunaga, K.: "Design of Multistory Settling Tanks in Osaka," Jour. Water Poll. Control Fed., 52, 950 (1980).

Mendis, J. B., and Benedek, A.: "Tube Settlers in Secondary Clarification of Domestic Wastewater," Jour. Water Poll. Control Fed., 52, 1893 (1980).

Mulbarger, M. C., and Huffman, D. D.: "Mixed Liquor Solids Separation by Flotation," Jour. San. Eng. Div., Proc. Am. Soc. Civ. Eng., 96, SA4, 861 (1970).

Munch, W. L., and FitzPatrick, J. A.: "Performance of Circular Final Clarifiers of an Activated Sludge Plant," Jour. Water Poll. Control Fed., 50, 265 (1978).

Pflanz, P.: "Performance of (Activated Sludge) Secondary Sedimentation Basins," paper presented at the 4th Int. Conference on Water Poll. Research, Prague, Czechoslovakia, 1969.

Poduska, R. A., Hicks, J. E., and Howard, J. W.: "Polymer Addition for Effluent Quality Control," Jour. Water Poll. Control Fed., 51, 2615 (1979).

Ramagoli, R. J.: "Polyurethane Foam Filtration," Proc. 33rd Purdue Industrial Wastes Conference, Lafayette, Indiana, May, 1978.

Schaffner, M. W., and Pipes, W. O.: "Underflow Rate and Control of an Activated Sludge Process," Jour. Water Poll. Control Fed., 50, 20 (1978).

Slechta, A. F., and Conley, W. R.: "Recent Experiences in Plant Scale Applications of the Settling Tube Concept," Jour. Water Poll. Control Fed., 43, 1724 (1971).

Swanson, C. L., and FitzPatrick, J. A.: "Comparison of Several Tertiary Wastewater Filter Designs," Proc. 33rd Purdue Industrial Wastes Conference, Lafayette, Indiana, May 1978.

Task Committee on Final Clarification: "Final Clarifiers for Activated Sludge Plants," Proc. Am. Soc. Civil Engrs., Jour. Envir. Eng. Div., 105, EE5, 803 (1979).

Tchobanoglous, G.: "Filtration Techniques in Tertiary Treatment," Jour. Water Poll. Control Fed., 42, 604 (1970).

Villemonte, J. R., and Rohlich, G. A.: "Hydraulic Characteristics of Circular Sedimentation Basins," Proc. 17th Purdue Industrial Waste Conf., Lafayette, Indiana, 682 (1963).

Wood, R. F., and Dick, R. I.: "Factors Influencing Batch Flotation Tests," Jour. Water Poll. Control Fed., 39, 304 (1973).

Dewatering, Utilization, and Disposal of Excess Activated Sludge

7.1 PROPERTIES OF ACTIVATED SLUDGE

The amounts of excess activated sludge depend on the type of wastewater treated, the application and effectiveness of wastewater clarification prior to activated sludge treatment, the selection of a treatment process modification, and the selection of the treatment parameters (see Chapters II, III, and IV). The excess of activated sludge is produced in the form of a liquid which usually contains prior to thickening (see Chapter VI), from 0.2 to 2 percent of solids which are volatile in 60-80 percent. Sludges from sewage treatment contain 2.5-5 percent of nitrogen, and 3-6 percent of P_2O_5. Heavy metal contents of activated sludges vary considerably. The average heating value of activated sludge dry solids is about 10,000 BTU per lb (5,550 cal per g).

The two most critical factors in the disposal or utilization of excess activated sludge are the sludge thickening and dewatering abilities. The success of the sludge thickening and dewatering operations determines the volume and/or energy requirements of the subsequent sludge handling units. The dewatering properties of sludges may be studied by determining their capillary suction time (Baskerville and Gale, 1968), specific resistance to filtration (Coackley and Jones, 1956; and Swanwick and Davidson, 1961), and coefficient of compressibility (Coackley and Jones, 1956). All these factors are influenced by the distribution of the sludge particle sizes (Karr and Keinath, 1978). For the most effective vacuum filtration of the excess activated sludge, filter leaf tests are run to determine the best cloth and the best chemical additive, and its dose.

The cost of the excess activated sludge handling and ultimate disposal is a substantial part of the total wastewater treatment cost. In sewage treatment plants, this sludge, called "secondary sludge," is usually treated together with the "primary sludge," which is separated from the sewage in

the primary clarifiers. In many treatment plants for industrial effluents,
the secondary sludge is the only sludge produced. In these cases, there
was a trend—until recently—towards the application of various "natural"
disposal schemes, such as land irrigation, controlled storage, ocean dump-
ing, etc. These possibilities have been severely limited by environmental
legislature.

In this Chapter, sludge dewatering by filtration will be described, followed by sludge thermal conditioning, stabilization by anaerobic and aerobic digestion, wet oxidation, conversion into animal food and other useful
materials and disposal by irrigation.

7.2 SLUDGE DEWATERING BY FILTRATION

Waste activated sludge dewatering by filtration can be accomplished by the
application of vacuum filters, filter presses, and belt filter presses.

VACUUM FILTRATION

Although properly operating activated sludge processes develop well-floccu-
lated sludge through the biological production of polymeric flocculants,
these naturally-produced materials are not effective enough as conditioners
of waste sludge for dewatering by vacuum filtration. For this purpose, the
addition of other "diluting" materials, or chemicals acting as polyelectro-

TABLE 7-1

Typical Waste Activated Sludge Dewatering on Vacuum Filters
with Cloth Media or in Filter Plate Presses
(after Environmental Protection Agency, 1979)

Factors	Vacuum Filters	Filter Plate Presses
Feed Solids Concentration, %	2.5-4.5	1.0-5.0
Chemical Dosage, lbs/ton dry solids		
$FeCl_3$	120-200	150
CaO	240-360	300
Yield, lb dry solids/sq ft/hr	1.0-3.0	—
Cycle Time, hrs.	—	2.5
Cake Concentration, %	13-20	45

Figure 7-1. Typical vacuum filter for sludge dewatering (photo author).

lytes, is required (Tenney and Cole, 1968; Tenney et al., 1970; Novak and
Haugan, 1980; etc.). In general activated sludge dewaters poorly through
vacuum filtration (Table 7-1). Any excessive presence of filamentous
growth in the sludge, and/or existence of dispersed and pinpoint flocs,
further worsen the effectiveness of the operation.

Figure 7-1 presents a typical vacuum filter for sludge dewatering.
Performance of such units is formulated as "yield" (capacity in terms of
dry solids mass per surface unit in a unit of time) and as the concentration
of the received cake (in percent of dry solids). Both performance measures
can be improved by increasing the solids concentration in the feed, and by
better sludge chemical conditioning.

FILTER PLATE PRESSES

Activated sludge dewatering in filter plate presses is capable of producing
highly concentrated cakes, but before the introduction of extensive
automation of the operation, it was not very popular, due to high labour
requirements and the rather unpleasant type of work required. Graefen
and Donges (1970) studied the parameters affecting sludge dewatering in
filter presses. A schematic of this operation is presented in Fig. 7-2,
and a typical example of the performance of filter plate presses dewatering
of waste activated sludge is given in Table 7-1.

BELT FILTER PRESSES

The application of belt filter presses is a relatively new approach to dewatering of excess activated sludge. Belt filter presses consist of one or more endless or seamed belts upon which sludge is deposited and progressively dewatered to a cake (Fig. 7-3). This system usually operates in three sequential stages of sludge dewatering: gravity drainage, low (wedge) pressure zone, and "high" (roll) pressure zone (shear zone). In each operating zone, sludge loses a part of its water content and increases its strength for treatment in the subsequent zone. The rates of increase of applied pressures in low pressure and high pressure zones differ markedly in the various belt filter presses presently marketed, and this limits their application for some types of sludge. Usually a polymer conditioning is required for effective dewatering of activated sludge on belt filter presses. Such conditioning may cost up to $50 and more per ton of activated sludge solids. The concentration of cake which can be achieved is 10 to 20% solids. Depending on the specific characteristics of the sludge, a tradeoff between belt filter press throughput and required sludge conditioning can be established. Solids recovery and cake concentration are also factors in the determination of the conditioning requirements.

The tightness of the filter belts affects the solids recovery efficiency and the possibility of producing a more concentrated cake. Typical tighter media are of 12 to 25 threads per cm and those generally used are of 6 to 8 threads per cm. There are three main causes which can shorten belt life: (1) puncturing, (2) abrasion and wear, and (3) failures associated with seamed belts. A properly operated filter belt will last from 3 to 4 months. An exchange of belt is a simple operation.

7.3 THERMAL CONDITIONING

To improve the dewatering performance of excess activated sludge, thermal conditioning of this material may be applied. It can be heating or freezing of the sludge.

Heat conditioning of activated sludge, often called the Porteous process, improves effectively the dewatering characteristics of this material. Usually this process is applied to the mixed sludges (primary and secondary) which are preheated to about 350°F (175°C), and passed to the reactor, along with steam, at up to 1,500 psi (100 atm). They are retained in the reactor for approximately 30 min. at temperatures of 350-390°F (180-200°C). The product of the treatment is used in a heat exchanger to heat the incoming sludge, and then is thickened by gravity and filtered or centrifuged. Heat treatment results in a 35-50 percent reduction in the volume of the activated sludge. The supernatant has a high BOD and should be biologically treated. Basic research on heat treatment of activated sludge was done by Brooks (1968), who later extended his research

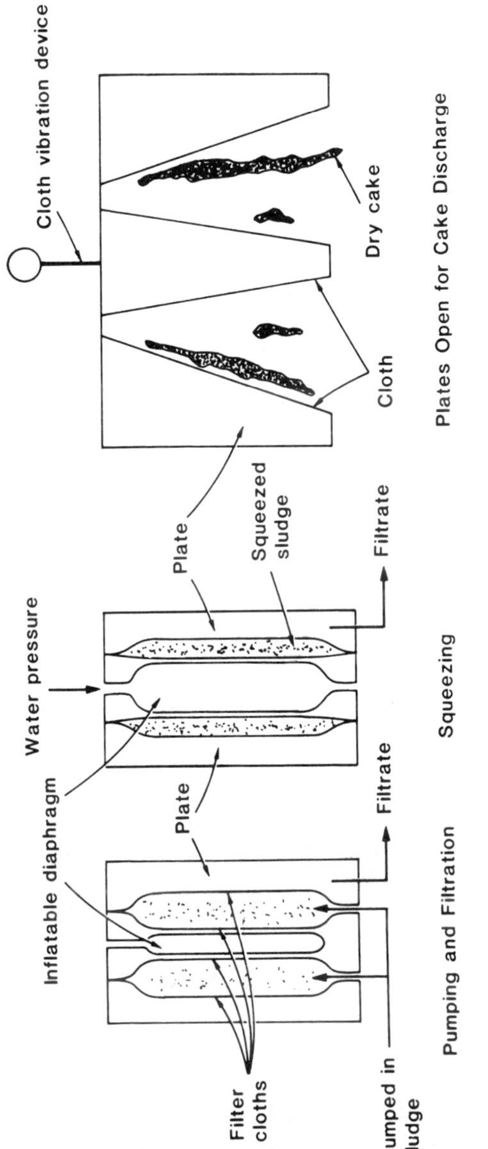

Figure 7-2. Schematic presentation of filter plate presses operation.

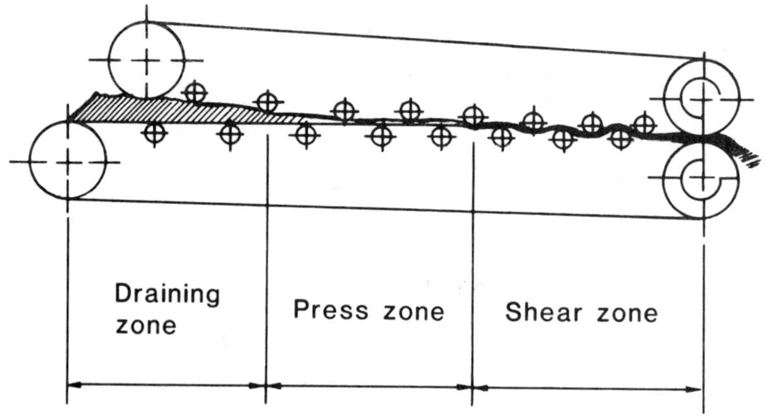

Figure 7-3. Schematic presentation of filter belt presses operation.

studies to other sewage sludges (1970). This method is based simply on
pressure cooking of the sludge. In such a treatment, the structure of
sludge is changed, its specific resistance to filtration is reduced and a
part of the suspended solids is solubilized. The mechanism of this process is not fully explained yet, but it is generally accepted that a hydrolysis
of some organic macromolecules is its main part. The products of the
hydrolysis: amino acids, volatile acids, simple sugars, etc. are responsible for the high BOD and COD of the resulting supernatant liquor. The
effect of heat pretreatment on anaerobic digestibility and dewaterability
of waste activated sludge and mixed sludges was studied by Haug et al.
(1978). It was found that, for waste activated sludge, heat pretreatment
could result in an increase (up to 60%) of methane gas production from digestion. The optimal temperature for such pretreatment was 175°C, but
heat pretreatment at temperatures over 200°C caused reduced gas production.

Indirect freezing of waste activated sludge was studied in a laboratory
scale by Katz and Mason (1970). They found that activated sludge conditioned this way could be dewatered by gravity draining on 40- to 80-mesh
screens to achieve a cake, with a solids concentration comparable to that
obtained by vacuum filtration after chemical conditioning. However, the
yield was 10-20 times that normally achieved in the latter case. The
mechanism of conditioning by slow freezing was explained by dehydration
and pressure effects. Direct freezing of waste activated sludge by liquid
butane was studied by Ali Khan et al. (1974). It was found that the sludge
handling and dewatering characteristics were a function of the freezing
time. Filtration rates were improved by 350 percent; drainage rates were

improved by as much as 1000 percent; and sludge moisture contents after drainage were frequently less than 50 percent of the original values. Incineration of a vacuum filtration cake produced from such sludge could be a self-sustaining process. Natural sludge refrigeration in Alaska was studied by Tilsworth et al. (1972).

7.4 AEROBIC AND ANAEROBIC DIGESTION

Biological stabilization of waste activated sludge can be accomplished either by aerobic or by anaerobic treatment. Generally, aerobic treatment produces stabilized material which is more difficult to dewater. It is also a more energy intensive process. However, it is less sensitive to the inhibitory effects of some sludge constituents (for example, heavy metals), produces relatively weak supernatant, and the facilities are less expensive and relatively easy to operate. On the other hand, anaerobic treatment produces material which is relatively easy to dewater and it may consume relatively little heat energy (or even produce some excess of energy). It does, however, solubilize a larger fraction of the sludge, producing a concentrated supernatant, is very sensitive to various inhibitory or toxic agents, and the more expensive facilities require more skilled operators. Economical comparisons between these two processes (for example, by Hogg and Ganczarczyk, 1973) depend to a great extent on the size of the required facility and on local conditions. However, recent increases in energy costs are making the anaerobic process more attractive in a large number of possible applications.

AEROBIC DIGESTION

In the process of aerobic digestion, excess activated sludge is aerated for an extended period of time to decompose its biodegradable fraction. This process is connected with some release of water bound by the sludge. The product of such treatment is characterized by a relatively low oxygen uptake rate, and dewaters easily enough to satisfy requirements for land disposal. Recommended loadings and oxygen requirements for the aerobic digestion of waste activated sludge are presented in Table 7-2.

In general, aerobic sludge digestion is suggested for small plants, and the key issue for the success of treatment is effective sludge thickening (Ganczarczyk et al., 1980) before aerobic digestion or during aerobic digestion. The process is sensitive to decreases in temperature (Koers and Mavinic, 1977; and Mavinic and Koers, 1979). An application of commercial oxygen instead of air for this process appears advantageous (Cohen, 1977). Unwanted nitrification during the aerobic digestion process may be controlled by substrate concentration (Hamoda and Ganczarczyk, 1980). The addition of chemicals may be used to improve dewatering of aerobically

TABLE 7-2

Requirements for Aerobic Digestion of Waste Activated Sludge

PARAMETER	VALUE	REMARKS
Additional solids retention time (sludge age), days	10-100	depending on the type of sludge, and the process temperature
Minimum oxygenation capacity, gO_2 per g MLVSS decomposed	2	enough air, or mechanical aerators power, to maintain a minimal dissolved oxygen level at 2 mg/L, and to keep the solids in suspension
Volatile suspended solids reduction, percent	0-50	depending on the type of sludge, and the composition of mineral solids in the sludge

digested sludges. The released water (supernatant) will pool over the sludge in sludge lagoons or ponds, and should be drained off quickly (Fig. 7-4).

AUTOHEATED AEROBIC DIGESTION

Aerobic digestion of waste activated sludge is basically an exothermic reaction. However, due to water evaporation from normal process facilities, the temperature increases observed are relatively low as the heat produced may not significantly outweigh the heat losses. By using covered reactors and oxygen instead of air (Matsch and Drnevich, 1977), the process can be operated even at the thermophilic temperature range (45-65°C), without any additional heat source. Similar effects were also achieved by Jewell and Kabrick (1980), using special self-aspirating air aerators.

ANAEROBIC DIGESTION

Only in relatively few situations (like the treatment of some industrial effluents), is an excess of activated sludge the only substrate for anaerobic digestion. Usually, the activated sludge is being aerobically digested as a mixture with primary sludge. Digestion of excess sludge from the contact stabilization plants which often operate without any primary clarification, is almost like the treatment of mixed primary and secondary sludges. Anaerobic digestion of organic sludges is generally a two-phase process. In the first phase, complex organic substances are converted into simple organic acids. In the second phase of the process, the simple organic

Figure 7-4. Dried, aerobically-stabilized sludge in sludge lagoons of the Dunville, Ontario, Sewage Treatment Plant (photo author).

acids are converted into methane and carbon dioxide. In current engineering practice, both stages of the process are not separated, although such a
possibility does exist, and has been the subject of several scientific studies
(e.g. Ghosh et al., 1975). The process technology is controlled by temperature, pH, and the presence of inhibitory substances like some heavy
metals which may accumulate in the sludge. In the process design, consideration should be given to the adequate mixing and heating of the system.
The anaerobic digesters are designed for one-stage treatment or for two-
stage treatment.

The requirements for anaerobic digestion of mixed sludge (primary and
secondary) have been presented in a detailed way by the Water Pollution
Control Federation (1977), and the Environmental Protection Agency (1979).

Anaerobic digestion of activated sludge only, is a more difficult process, and usually requires a decreased loading rate compared with the
above data. In comparison with anaerobic digestion of primary solids, it
leads to smaller decreases in organic solids (up to by 30-35%), produces
only about 50% of the fermentation gas per unit of decomposed volatile
solids and produces digested sludge with a limited ability to dewater.

Supernatant from anaerobic digestion of activated sludge contains considerable amounts of dissolved and colloidal organic substances, suspended
solids, and soluble nitrogen and phosphorus compounds. Qualitative characteristics of the supernatant from anaerobic digestion of mixed primary
and waste activated sludges are presented in Table 7-3. The quantity of
anaerobic supernatant depends on several factors affecting the sludge characteristics and the anaerobic digestion process.

Anaerobic digestion of the excess activated sludge from the treatment
of some industrial effluents may not be effective enough to suggest this
treatment process at all.

TABLE 7-3

Supernatant Characteristics from Anaerobic Digestion
of Mixed Primary Sludge and Waste Activated Sludge
(after Kappe, 1958)

Analytical Determination	Avg.	Max.	Min.
Suspended solids, mg/L	4,408	14,650	100
Volatile suspended solids, mg/L	3,176	10,650	75
BOD, mg/L O_2	667	2,700	100

AEROBIC/ANAEROBIC DIGESTION

A new Union Carbide combined process for sludge digestion consists of a
short, aerobic oxygen thermophilic digestion system (for example, at
1.2 days retention), followed by high rate anaerobic digestion facilities.
It is claimed that such an approach eliminates weaknesses in both aerobic
and anaerobic processes and enhances their benefits. This new process
was first demonstrated at Hagerstown, Maryland, and has been studied at
Lackawanna, New York, and funded by U.S. Environmental Protection
Agency as innovative technology.

7.5 WET AIR OXIDATION AND PYROLYSIS OF SLUDGE

The wet oxidation process for treatment of organic sludges is often referred to as the Zimmermann Process or Zimpro. This process is based
on air oxidation of a part of the sludge organic material, at increased temperatures and pressures (Hurwitz et al., 1965). These two parameters
govern both the time of treatment and the degree of sludge oxidation. The
heat released by oxidation of organics supports the temperature of the reaction, and can be utilized for the power generation required to compress
the air introduced to the system. Figure 7-5 shows a typical flowsheet of
the wet oxidation process. Preheated sludge and air are forced into the
reactor by high pressure pumps. The products of the oxidation reaction
are separated into gaseous and liquid phases. The gases (steam, carbon
dioxide, nitrogen and excess of oxygen) are used for power generation.
The hot liquid suspension is used to preheat the incoming sludge in a heat
exchanger. The wet oxidation process usually operates at temperatures in
the range of 480–570°F (250–300°C) and pressures in the range of 1,000 to
2,200 psi (68–150 atm).

Wet air oxidation drastically improves the thickening and dewatering
properties of sludges. Even oxidation in the order of 15 to 20 percent of
the organic matter in a sludge causes rapid draining and good filterability
of activated sludges. The supernatant obtained in wet oxidation of activated
sludges contains dissolved and colloidal organic substances and should be
directed for biological treatment.

The process of wet oxidation of activated sludges is expensive because
of high capital, maintenance (corrosion problems) and operational (energy
consumption) costs. Economical application of this process can only be
expected at very large installations where good heat recovery and generation of additional power are possible. Another problem is the odours associated with such treatment.

A mathematical model for the wet-air oxidation of sludge was presented by Takamatsu et al. (1970), and the effects of wet-air oxidation
were studied by Sommers and Curtis (1977). It was found that this

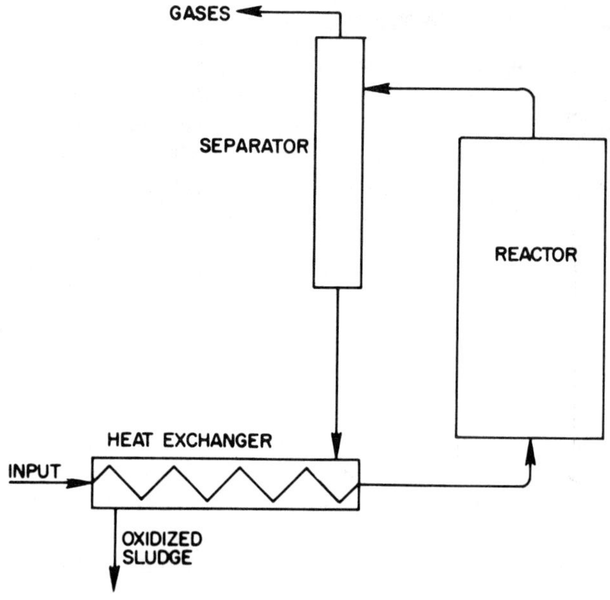

Figure 7-5. Flowsheet of waste activated sludge wet-oxidation.

process drastically decreased the nitrogen contents of sludge, and, therefore, decreased its value as a fertilizer.

An alternative to wet oxidation of excess activated sludge may be its
pyrolysis (destruction by heat) in either an oxygen-free or an oxygen deficient atmosphere. The products of such treatment are combustible gases,
some oils and carbon char. Srinivasaraghavan et al. (1976) reviewed the
state of this process development.

7.6 SLUDGE DRYING AND INCINERATION

Waste activated sludge, after thickening and dewatering, may be dried in
different ways. Biologically stabilized excess activated sludge may be
dried on sand drying beds or in sludge lagoons (see Fig. 7-4) under proper
climatic conditions. The usual requirement is to apply only thin layers of
thickened sludge (up to 4 in) on the beds. Once the sludge is dried, it will
not get wet very easily.

One of the recently recommended drying methods for activated sludge
is the Carver-Greenfield Process (Bress, 1979) which is based on the conversion of wet sludge into hydrocarbon oil slurry and the carrying of this
slurry through a multi-effect evaporation system.

The control of emissions and odours is a necessary part of the incineration process for waste activated sludge alone or with primary sludge. The use of exhaust gas scrubbers or afterburners is advisable. A well documented review of waste activated sludge incineration case histories was presented by Wells et al. (1979).

7.7 SLUDGE UTILIZATION

Waste activated sludge can be utilized for chemical processing, conversion to animal food, and land application as a fertilizer and/or soil conditioner. None of these potential applications is generally economically attractive at the present time. However, some individual solutions, especially for sludges from some industrial installations, do have merits.

CHEMICAL PROCESSING

Several efforts to convert waste activated sludge into plastic materials were not successful. Activated carbon with a surface of about 4,300 sq. ft. (400 sq. m) per g of carbon may be prepared from activated sludge (Bosh et al., 1976) by carbonization and activation with steam at 1,500–1,640°F (800–900°C). The inorganic matter in the sludge reduces the active surface area to 2,150 sq. ft. (200 sq. m) per g of the product. This activated carbon can be used effectively for the activated carbon–activated sludge treatment process (see Chapter IV, p. 4.11).

ANIMAL FOOD PRODUCTION

The 20 to 60% protein fraction of activated sludge can make this material a valuable supplement to various types of animal feed, directly after heat drying or other effective dewatering procedures such as solvent extraction. In the past, considerable interest was shown in this use for excess activated sludge (Hurwitz, 1957; Hackler et al., 1957), but the health aspects of this material seriously limit its application. The possibility of active pathogenic bacteria and viruses being present in the sludge is suspected, and possible fungal content may be associated with the presence of directly toxic organic substances. Also heavy metals may find a route to the food cycle in this way. Therefore, only a very small percentage of this supplement may be acceptable in animal foods. An exception to this rule may be sludge from the treatment of wastewater which, itself, contains valuable proteins, such as the wastewater from canneries (Esvelt et al., 1976), meat packing (Paulson and Lively, 1979), etc.

An alternative to direct activated sludge utilization for animal food production is an indirect method by acid hydrolysis of the sludge and the production in this way of a concentrated extract (Bouthilet and Dean, 1970).

TABLE 7-4
Major Constituents of Milorganite
(after Sewerage Commission of the City
of Milwaukee, 1968)

Organic matter, percent	75.00
Organic nitrogen, percent	6.26
Water insoluble nitrogen, percent	6.02
Total phosphoric acid, percent	4.15
Available phosphoric acid, percent	3.40
Potassium as K_2O, percent	0.50
Iron as Fe, percent	3.59
Sulphur as SO_3, percent	2.79
Magnesium as MgO, percent	1.74
Calcium as CaO, percent	1.61
Copper	
Manganese	
Zinc	Present
Molybdenum	
Titanium	
Nickel	

PRODUCTION OF FERTILIZERS AND SOIL CONDITIONERS

Although waste activated sludge is not a very good fertilizer, it has some
soil conditioning properties, and can be conveniently applied directly on
land wherever heavy metals pollution of crops is not a problem (non-food
crops or sludges poor in heavy metals). An example of the fertilizing of
a Kentucky Bluegrass crop with excess activated sludge from the treatment
of sulfite pulping wastewater is described by Eberhardt et al. (1978).

Various types of plant fertilizers and soil conditioners can be produced
from the excess activated sludge in a dry form. The excess activated
sludge from Milwaukee, Wisconsin, Jones Island Treatment Plant is converted into a fertilizer called Milorganite. This plant treats screened
sewage with 4-8 hrs. aeration, with 25-35 percent of the recycling sludge.
The excess sludge is thickened, conditioned with ferric chloride, filtered
on vacuum filters and dried in rotary dryers. Approximately 75,000 tons
of Milorganite (Table 7-4) were produced and sold yearly until the question

of heavy metals pollution was raised. At present, this material is used
only on golf courses and for flowers.

In a similar way, part of the excess activated sludge from the West-
Southwest Plant of Metropolitan Chicago is used for the production of a
fertilizer. In this plant, sewage is treated by short-time primary clarification, and 3-1/2 hrs. aeration, with 30 percent of return sludge. The
activated sludge used for the fertilizer production is thickened for
4-1/2 hrs., conditioned with ferric chloride, vacuum filtered and heat
dried.

REFERENCES

Ali Khan, M. Z., et al.: "The Conditioning of Waste Activated Sludge by
the Direct Slurry Freezing Process," paper presented at the 47th Annual
Conference of Water Pollution Control Federation, Denver, Colorado,
October, 1974.

Baskerville, R. C., and Gale, R. S.: "A Simple Automatic Instrument
for Determining the Filterability of Sewage Sludges," Wat. Pollut. Control
(Brit.), 67, 233 (1968).

Bosh, H., et al.: "Activated Carbon from Activated Sludge," Jour. Water
Poll. Control Fed., 48, 551 (1976).

Bouthilet, R. J., and Dean, R. B.: "Hydrolysis of Activated Sludge,"
paper presented at the 5th Int. Water Poll. Research Conf., San Francisco,
1970.

Bress, D. F.: "Carver-Greenfield Process Drying of Pulp Mill Activated
Sludge for Energy Recovery in a Hog Fuel Boiler," paper presented at the
34th Ann. Purdue Industrial Waste Conference, May, 1979.

Brooks, R. B.: "Heat Treatment of Activated Sludge," Water Pollution
Control, 67, 592 (1968).

Brooks, R. B.: "Heat Treatment of Sewage Sludges," Water Pollution Control (Brit.), 67, 92, 221 (1970).

Coackley, P. C., and Jones, B. R. S.: "Vacuum Sludge Filtration Interpretation of Results by the Concept of Specific Resistance," Sewage and
Ind. Wastes, 28, 963 (1956).

Cohen, D.: "A Comparison of Pure Oxygen and Diffused Air Digestion of
Waste Activated Sludge," Prog. Water Techn., 9, 691 (1977).

Eberhart, W. A., et al: "Conversion of Sulfite Pulping Waste Sludge to
an Agricultural Product," Jour. Water Poll. Control Fed., 50, 1893
(1978).

Esvelt, L. A., et al.: "Cannery Waste Biological Sludge Disposal as Cattle Feed," Jour. Water Poll. Control Fed., 48, 2778 (1976).

Environmental Protection Agency: "Process Design Manual for Sludge Treatment and Disposal," 625/1-79-011 (1979).

Ganczarczyk, J., et al.: "Performance of Aerobic Digestion at Different Sludge Solid Levels and Operation Patterns," Water Research (Brit.), 14, 627, (1980).

Ghosh, S., et al.: "Anaerobic Acidogenesis of Wastewater Sludge," Jour. Water Poll. Control Fed., 47, 30 (1975).

Graefen, H., and Donges, H. J.: "Studies on Parameters Affecting Sludge Dewatering in Pressure Filters," paper presented at the 5th Conference of Int. Assoc. on Water Poll. Research, San Francisco, 1970.

Hackler, L. R., et al.: "Feed from Sewage. III. Dried Activated Sewage Sludge as a Nitrogen Source for Sheep," Jour. Animal Sci., 16, 125 (1957).

Hamoda, M. F., and Ganczarczyk, J.: "Control of Aerobic Digestion by Substrate Concentration," Can. Jour. Civil Eng., 7, 456 (1980).

Haug, R. T., et al.: "Effect of Thermal Pretreatment on Digestability and Dewaterability of Organic Sludges," Jour. Water Poll. Control Fed., 50, 73 (1978).

Hogg, K. S., and Ganczarczyk, J.: "A Performance Study of Aerobic Sludge Digestion," Proc. Water Pollution Research in Canada, 8, 26, (1973).

Hurwitz, E. et al.: "Wet-Air Oxidation of Sewage Sludge," Water & Sewage Wks., 112, 293 (1965).

Hurwitz, R.: "Feeding Dried Activated Sludge to Pigs, Poultry, Steers, and Sheep," Wastes Eng., 28, 399 (1957).

Jewell, W. J., and Kabrick, R. M.: "Autoheated Aerobic Thermophilic Digestion with Aeration," Jour. Water Poll. Control Fed., 52, 512 (1980).

Kappe, S. E.: "Digester Supernatant: Problems, Characteristics and Treatment," Sewage and Ind. Waste, 30, 937 (1958).

Karr, P. R., and Keinath, T. M.: "Influence of Particle Size on Sludge Dewaterability," Jour. Water Poll. Control Fed., 50, 1911 (1978).

Katz, W. J., and Mason, D. G.: "Freezing Methods Used to Condition Activated Sludge," Water & Sewage Wks., 117, 110 (1970).

Koers, D. A., and Mavinic, D. S.: "Aerobic Digestion of Waste Activated Sludge at Low Temperatures," Jour. Water Poll. Control Fed., 49, 460 (1977).

Matsch, L. C. and Drnevich, R. F.: "Autothermal Aerobic Digestion," Jour. Water Poll. Control Fed., 49, 296 (1977).

Mavinic, D. S., and Koers, D. A.: "Performance and Kinetics of Low-Temperature Aerobic Sludge Digestion," Jour. Water Poll. Control Fed., 51, 2088 (1979).

Novak, J. T., and Haugan, B. E.: "Mechanisms and Methods for Polymer Conditioning of Activated Sludge," Jour. Water Poll. Control Fed., 52, 2571 (1980).

Paulson, W. L., and Lively, L. D.: "Oxidation Ditch Treatment of Meat-packing Wastes," EPA-600/2-79-030, January, 1979.

Sewerage Commission of the City of Milwaukee: "Milwaukee Waste Water Treatment Facilities," Milwaukee, Wisconsin, 1968.

Sommers, L. E., and Curtis, E. H.: "Wet-Air Oxidation: Effect on Sludge Composition," Jour. Water Poll. Control Fed., 49, 2219 (1977).

Srinivasaraghavan, R. et al.: "Evaluation of Pyrolysis as a Wastewater Sludge Disposal Alternative," paper presented at the 49th Annual Conference of the Water Pollution Control Federation, Minneapolis, Minnesota, 1976.

Swanwick, J. D., and Davidson, M. F.: "Determination of Specific Resistance to Filtration," Wat. Waste Treatm. (Brit.) 8, 386 (1961).

Takamatsu, T. et al.: "A Mathematical Model for the Wet-Air Oxidation of Sludge", paper presented at the 5th Int. Water Poll. Research Conf., San Francisco, 1970.

Tenney, M. W. and Cole, T. G.: "The Use of Fly Ash in Conditioning Biological Sludges for Vacuum Filtration," Jour. Water Poll. Control Fed., 40, R281 (1968).

Tenney, M. W., et al.: "Chemical Conditioning of Biological Sludge for Vacuum Filtration," Jour. Water Poll. Control Fed., 42, 212, R1 (1970).

Tilsworth, T., et al.: "Freeze Conditioning of Waste Activated Sludge," paper presented at the 27th Purdue Industrial Waste Conference, Lafayette, Indiana, 1972.

Water Pollution Control Federation: "Wastewater Treatment Plant Design," Manual of Practice No. 8, Washington, D.C., 1977.

Water Pollution Control Federation: "Sludge Thickening," Manual of Practice No. FD-1, Washington, D.C., 1980.

Wells, J. F. et al.: "Case Histories of Waste Activated Sludge Incineration," Jour. Water Poll. Control Fed., 51, 2886 (1979).

Application of Activated Sludge Process to Industrial Wastewater Treatment

8.1 REQUIREMENTS FOR TREATMENT

Although the activated sludge process originally was developed for municipal sewage treatment, even in the early stages of its application it was used for the treatment of municipal sewage which contained a substantial amount of industrial effluent and for the treatment of industrial effluents only. There are, however, several conditions which have to be fulfilled for a successful activated sludge treatment of industrial wastewater containing bio-degradable organic compounds. They include:

proper wastewater pre-treatment (see p. 8.2) to remove or inactivate toxic or inhibitory components, to make the wastewater more amenable to biological treatment, and/or to make such treatment more economical;

supplementation of the required mineral nutrients (see p. 2.2); and

adaptation of activated sludge micro-organisms to the specific metabolic substrates and, sometimes, to particular characteristics of the treatment system (like the presence of some inhibitory substances, low or high pH, high salinity, etc.).

ACTIVATED SLUDGE ADAPTATION

The possibilities of adapting activated sludge micro-organisms for the decomposition of organic substances which are drastically different from municipal sewage organic components (such as carbohydrates, fats, and proteins) are very good (Davies and Gale, 1953; Kabler, 1960; etc.). It is possible to biologically decompose several compounds which in the past were considered resistant or even toxic to bacterial metabolism, like some phenols (McKinney et al., 1956), formaldehyde (Gellman and Heukelekian,

1950), acetone, ethylene glycol, and other alcohols (McKinney and Jeris, 1955), aliphatic and aromatic amines (Malaney, 1960), acrylontrile and other nitriles, various intermediates of organic synthesis, as well as many detergents. However, some substances present in the effluents from chemical plants (e.g., nitro-compounds) may adversely affect biological wastewater treatment, slowing down the process kinetics or even causing its failure. Many such substances can be considered at least mildly inhibitory to biological treatment, and some organic substances, although not inhibitory, cannot be decomposed during activated sludge treatment. In the activated sludge process treatment, the periods of inadequate acclimation of micro-organisms may be characterized with a lower than expected nutrients uptakes and oxygenation requirements. In case of some industrial effluents, the complete acclimation may be achieved only after a relatively long period of the plant operation.

As a source of micro-organisms for activated sludge treatment of industrial effluents, activated sludge from municipal sewage treatment, sewage primary sludge, top soil, river or lake bottom deposits, etc., may be used. Acclimation of micro-organisms to a specific industrial wastewater, in general, means the development of a significant number of organisms capable of decomposing the compound or compounds in question. Often, single bacteria species cannot degrade some complex molecules. In such situations, several species acting in sequence are required. As it was mentioned before, for some applications, the process of acclimation may be lengthy. It may require the development of mutant bacteria strains, generation of specific enzymes, etc. If inhibitory or toxic effects are anticipated, acclimation should begin with a relatively low concentration of respective substances. Inhibition or toxicity of a specific substance towards an activated sludge system, should presumably be considered in terms of ratio between the inhibitory or toxic material and the active biomass. Moreover, inhibition or toxicity appears often to be a function of the treatment rate (applied food to micro-organisms ratio per day) and increases drastically at higher BOD loading values. Therefore, it is difficult to determine, in general, a significant toxic concentration of a specific substance in wastewater.

An improved biodegradation of some organic components of industrial wastewater can be achieved by certain kinds of destructive oxidation pretreatment like hypochlorination, ozonation, etc. Such an "induced" biodegradability of sulfite pulp mill effluents was studied by Ganczarczyk et al. (1974). Hypochlorination and chlorine dioxide treatment of lignosulfonic acids in sulfite effluents caused the depolymerization of this material to a product containing from 5 to 15 lignin monomer units (Fig. 8-1). High molecular weight lignosulfonic acids composed of 250 to 550 monomer units were the most sensitive to such depolymerization treatment, and partial by-product of the depolymerization reaction, lignosulfonates of molecular weight in the range from 2,000 to 4,000 proved to be resistant both to biological treatment and to chemical oxidation. Narkis and

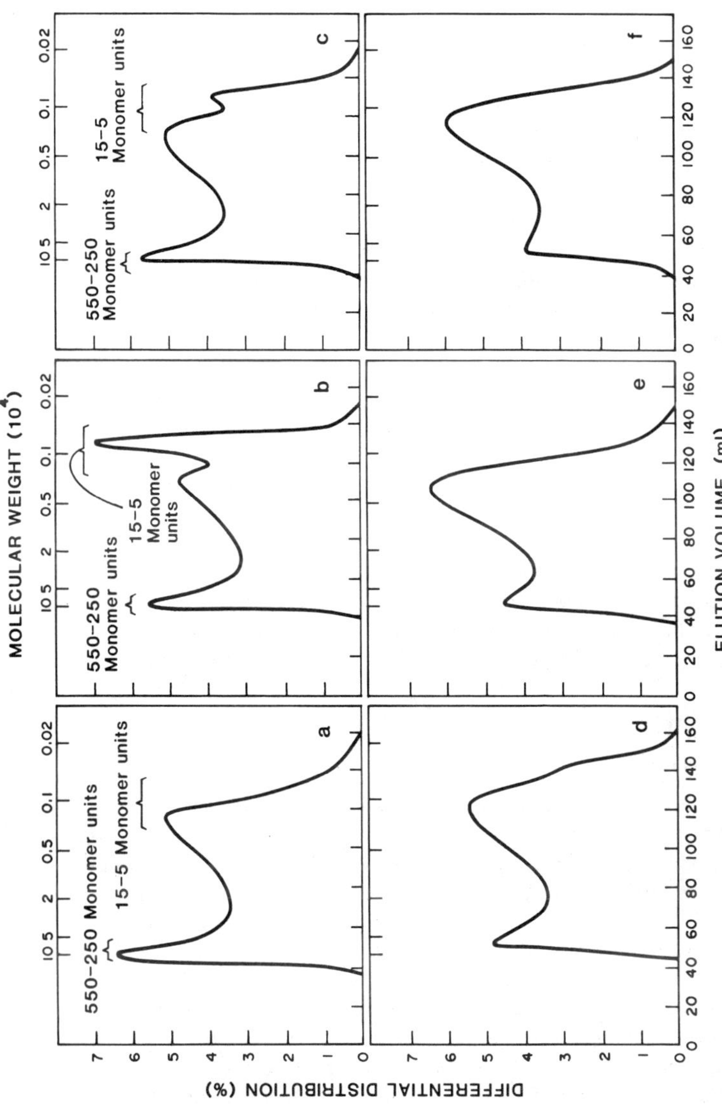

Figure 8-1. Changes in molecular weight distribution of lignosulfonic acids during hypochlorination of simulated sulfite effluents—active chlorine dosage: 1.17 mg/mg COD (a—untreated effluent, b—effluent immediately after chemical addition, c—after 12 hours, d—after 24 hours, e—after 36 hours, and f—after 48 hours).

Schneider-Rotel (1980) found that ozonation improved biodegradation of a
non-ionic surfactant (branched chain nonyl-phenolethoxylate) to the removal
of 70% COD in comparison with the removal of only 8-25% COD without
ozonation. This observation was explained by changes in the molecular
structure of the substance studied which were caused by the pre-treatment.

An effective biodegradation of organic chemicals present in industrial
wastewater can be characterized by acceptable biodegradation rates, which
can be defined in mass terms of organic substances eliminated per unit
mass of bacterial growth per unit of time, or as the half-life of the sum of
concentrations of the original organic compounds and their organic intermediates in the water phase. In the batch type studies, biodegradation in
general may start without delay, with some acceleration period, or with
some lag periods. Long lag periods are proof that a special adaptation of
microorganisms is required for degradation of the specific organic substance. Biodegradation studies involve analytical determination of the
specific organic substances and their intermediates. Often an acceptable
substitute may be the determination of COD or TOC utilization in the
biological system.

The various tests and methods for the determination of biological degradability of organic substances have been discussed by Ludzack and
Ettinger (1963). One of the suggested procedures (Pitter, 1976) calls for
determination of the degree and the rate of biological degradation in batch
experiments with an adapted activated sludge, and the studied substance as
a sole source of organic carbon. In European studies on the biodegradability of organic substances, a bioassay test suggested by Husmann and
based on the continuous operation of an activated sludge model has achieved
some popularity. This test depends heavily on the operational parameters
of the applied model. Its performance is relatively labour-intensive and
expensive.

HEAVY METALS

Many heavy metals affect activated sludge performance only when present
in wastewater in an ionized form. At higher pH values they are inactive,
and to a certain degree may be removed from the treated effluents by
sorption on bio-mass. This removal can be correlated with an increase
in metal concentrations in the sludge, and may result in the production of
sludge which is more difficult to treat and not acceptable for land application (Mytelka et al., 1973; Oliver and Cosgrove, 1974; etc.). In the past,
considerable effort was exercised to determine the process limiting concentrations of particular metals in wastewater directed to biological treatment (Barth et al. 1965). It appears, however, that such limits depend
heavily on the process modification and numerous other factors which
should be individually analyzed. A detailed review of the mechanisms of
metal removal by activated sludge, and the special role of bacterial extracellular polymers in this phenomenon, was presented by Brown and Lester

(1979). In a continuation of this study (Sterritt and Lester, 1981), the influence of sludge age on heavy metal removal was investigated, and, for
a selected sludge age, a different metal affinity series to the sludge was
presented.

NEUTRALIZATION

A general optimum for aerobic bacterial growth is between 6.5 and 7.5 pH.
Although this optimum may be slightly different for particular industrial
effluents, chemical neutralization is usually required only for strongly
acidic or alkaline wastewater prior to its activated sludge treatment.
For moderately acidic wastewater, an activated sludge treatment is possible but the sludge may become rich in fungi and show some difficulties
in separation operations. Treatment of slightly alkaline wastewater may
be sometimes easier than of neutral industrial wastewater, as higher pH
values usually improve flocculation of the activated sludge. However,
these values should not be high enough to show serious inhibitory effects
to the bacterial growth and metabolism.

Under practical conditions, chemical neutralization of acidic wastewater does not necessarily have to be carried to the neutrality point prior
to activated sludge treatment. In general, it is satisfactory to neutralize
only the strong mineral acids component of the acidity. If biodegradable
organic acids are present in the wastewater, the biological decomposition
of this component is connected with the removal of the corresponding
acidity. E.g., in a chemical plant in Fukuroi, Japan, the wastewater contained some mineral acids and some biodegradable organic acids. The
chemical neutralization with a caustic soda solution was practised only
to pH = 5.0. Subsequent activated sludge treatment increased the wastewater pH to 8.0 because of decomposition of those organic acids. However, if biological treatment of an industrial effluent is connected with the
production of a strong mineral acid, as in the nitrification process, the
corresponding alkalinity of the wastewater is utilized and supplementation
may be required (see Chapter II, p. 2.8).

In activated sludge treatment of alkaline wastewater, the carbon
dioxide developed can be utilized for chemical neutralization. Eckenfelder
(1967) estimated, under conditions of complete mixing, a neutralization of
0.7 lbs of caustic alkalinity (as $CaCO_3$) for 1 lb COD decomposed, and
suggested the need for separate neutralization only for the possible surplus
of the caustic alkalinity. Ganczarczyk (1969) conducted two studies on
chemical neutralization in a full-scale activated sludge treatment of unbleached Kraft pulp mill effluent at a conventional range of loadings in
aeration tanks with limited longitudinal mixing. In the first study period
(Summer, 1967), the average pH of the wastewater decreased from
9.7 to 8.2, the average alkalinity of "p" of 3.4 me/L was completely removed and the average alkalinity "m" was decreased from 10.2 me/L to
8.4 me/L. In a second study period (Summer, 1968), the wastewaters

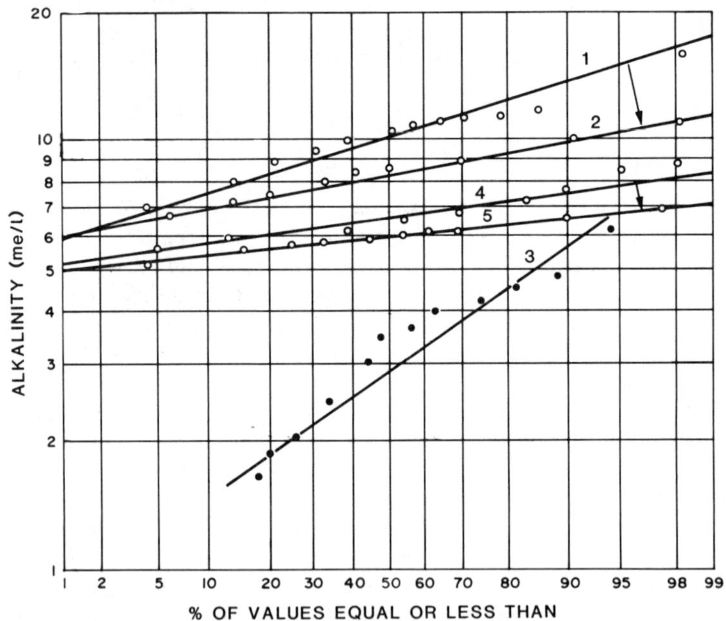

Figure 8-2. Distribution of wastewater alkalinity data (1—alkalinity "m"
of untreated wastewater in Period I, 2—alkalinity "m" of treated wastewater
in Period I, 3—alkalinity "p" of untreated wastewater in Period I, 4—alkalinity "m" of untreated wastewater in Period II, and 5—alkalinity "m" of
treated wastewater in Period II).

were substantially less alkaline. The average untreated wastewater pH
values were lower by 1.5 units (pH = 8.2) and those of the treated wastewater by 0.7 units (pH = 7.7). Simultaneously, the "p" alkalinity of the
untreated wastewater almost disappeared and the "m" alkalinity of the untreated and treated wastewater was substantially lower by 36.5 percent
(6.55 me/L) and 28.6 percent (6.0 me/L), respectively (Fig. 8-2). The
decreases in BOD and COD in both series of measurements were very
close to each other (see p. 8.5).

HIGH LEVELS OF DISSOLVED SOLIDS

The influence of high levels of dissolved solids in wastewater (salinity) on
activated sludge process performance is a function of the applied loads and
some other process parameters. For a conventional activated sludge
process, concentrations of chloride up to 5,000-8,000 mg/L usually are
not very effective, but at higher levels of salinity, process efficiency

steadily decreases and residual suspended solids increase, due to a decrease in the flocculation ability of the micro-organisms (Ludzack and Noran, 1965). Tokuz and Eckenfelder (1979) studied the influence of sodium chloride and sodium sulfate on the performance of the process. With thorough acclimation procedures, they achieved satisfactory BOD and suspended solids removals up to sodium chloride concentrations of 35,000 mg/L. However, the wastewater COD removals were affected by the increasing salinity. The effects of sodium sulfate on the systems studied were less profound than those of sodium chloride (see Chapter III, p. 3.10).

8.2 PRETREATMENT OF INDUSTRIAL EFFLUENTS

Pretreatment of industrial effluents prior to activated sludge treatment may cover changes in the physical characteristics of the wastewater (lack of uniformity of flow or composition, temperature, etc.) and/or the removal of substances which are not compatible with biological treatment or which are making such treatment more difficult or expensive. There are numerous technological unit operations and processes used for pretreatment of industrial effluents. Their scope ranges from simple equalization or sedimentation to chemical pre-oxidation with ozone or chlorine, selective carbon adsorption, anaerobic digestion, etc.

The objective of wastewater quantity and quality equalization is to dampen wastewater flow variations, to achieve a relatively uniform flow rate through aeration tanks and secondary clarifiers, and to provide possibly uniform organic loading of the activated sludge system. Such pretreatment can make possible treatment of some highly variable industrial effluents, and, in other cases, can improve the efficiency of the treatment and reduce the required size of the facilities (see Chapter III, p. 3.1). The design of equalization units depends on the acceptable range of wastewater uniformity. Equalization tanks can be designed as in-line or side-line basins with different degrees of compartmentalization, mixing and/or pumping equipment. Pre-aeration of wastewater is often applied as a mixing system in equalization facilities. It may be beneficial for flocculation of some suspended solids present in wastewater, may also contribute to removal by stripping of some volatile wastewater components, and provide an aerobic condition for the wastewater.

The preparation of coke-plant effluents for activated sludge dephenolization (Ganczarczyk, 1980) presents an interesting review of pretreatment strategy. There are several different wastewater streams in coke-plant effluents and the flow of particular streams may vary fairly widely. The chemical composition of the coal used for coking and the coal moisture content are the primary factors contributing to the characteristics of the collected effluents. They contain biodegradable and non-biodegradable organics, specific organics (like phenols), cyanide, thiocyanate, other mineral sulphur compounds and substantial concentrations of ammonia. In the majority of North American coke-plants, the effluents are composed of excess flushing liquor and condensates from primary and final coolers, and benzol plants. At some coke-plants, wastewaters from the ammonium sulphate crystallization, tar still, gas desulphurizer and cyanide stripper are connected to the above.

The usual technological methods for pretreatment of coke-plant effluents (Fig. 8-3) are as follows:

1. wastewater collection for retention, equalization, and tar separation;

2. flotation for removal of residual oils and tar;

3. phenol extraction;

   a. benzene-caustic method by Pott-Hilgenstock;

   b. phenosolvan method by Lurgi; etc. ;

4. ammonia stripping;

   a. free ammonia stripping;

   b. free and combined ammonia stripping following alkalization by addition of caustic soda;

   c. free and combined ammonia stripping following alkalization by addition of lime (the method requires clarification after stripping); and

5. dilution.

In North America, an acceptable system for the pretreatment of coke-plant effluents is most often composed of equalization and ammonia stripping (shown on Fig. 8-3 as darker rectangles). In Europe and in Japan, phenol extraction prior to ammonia stripping is often practised. In general, the selected pretreatment strategy should be a function of treatment

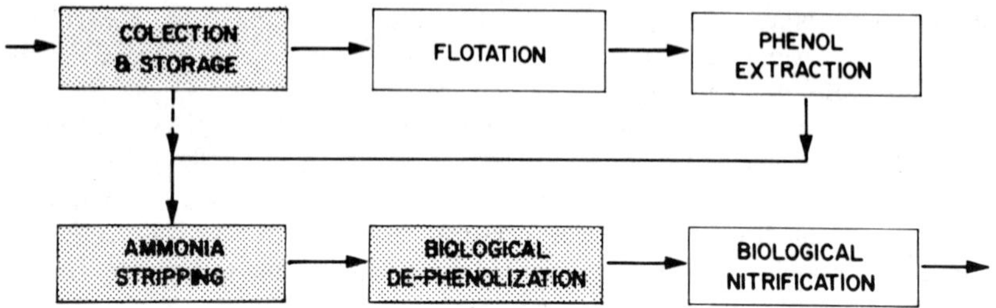

Figure 8-3. Typical pretreatment methods for coke-plant effluents.

requirements and process economy. However, not enough information is presently available to model the effectiveness of particular pretreatment steps and to optimize the whole system. At the various plants, the particular pretreatment units are designed differently and operate with widely varying degrees of effectiveness. Differences in the scope of the pretreatment methods practised are illustrated by the methods used in the following case examples:

1. Sumikin Coke Plant, Kashima (Japan):

    a. air flotation,

    b. phenosolvan extraction,

    c. stripping of free ammonia, and

    d. dilution;

2. Bethlehem Steel Plant, Bethlehem, Pennsylvania:

    a. storage, and

    b. dilution;

3. DOFASCO Plant, Hamilton, Ontario:

    a. storage and

    b. ammonia stripping with lime alkalization.

8.3 TREATMENT OF INDUSTRIAL EFFLUENTS WITH MUNICIPAL SEWAGE

Joint treatment of industrial effluents with municipal sewage, using the activated sludge process, is often technologically attractive and economically justified. This is due mostly to the larger scale of the joint treatment plant, allowing for better quality staff, and better instrumentation and control; capital costs savings are also achieved because of the scale of the plant itself, especially in the size range from 5-10 mgd (19,000-38,000 cu.m/d). Moreover, municipal sewage may be a satisfactory source of mineral nutrients for the joint wastewater treatment, and a source of micro-organisms continuously reseeding the activated sludge system. Municipal sewage may also dilute enough some inhibitory components in industrial effluents, making the joint treatment process much easier and more effective than would be possible for industrial effluents only.

On the other hand, the addition of some industrial effluents to municipal sewage may decrease the possibility of obtaining an effluent suitable for water recycling, recharging groundwater resources, etc. Groundwater infiltration in long gravity sewers required for some joint treatment cases,

TABLE 8-1
Inhibitory Threshold Concentration of Some Inorganic Components of Industrial Effluents (in mg/L)

Pollutants	Activated Sludge Process	Nitrification Process
Ammonia	480	
Arsenic	0.1	
Boron	0.05-100	
Cadmium	10-100	
Chromium (hexavalent)	1-10	0.25
Chromium (trivalent)	50	0.005-0.5
Copper	1.0	0.34
Cyanide	0.1-5	0.5
Lead	0.1	
Manganese	10	
Mercury	0.1-5.0	0.25
Nickel	1.0-2.5	
Zinc	0.08-10	0.08-0.5

may increase unnecessarily the flow to a centralized plant. In addition, to maintain stream quality, the high volume point discharge from a joint treatment plant must be of much better quality than the same total volume discharged at several different points. Also fluctuations in industrial production and industrial accidents may cause problems at the joint treatment plants. However, the magnitude of these problems is less pronounced than if they occurred in a treatment plant for industrial wastewater only.

The participation of industrial effluents in a joint treatment is often determined by so-called "population equivalents" for flow, BOD load, and suspended solids load. As these values differ considerably in different countries and under different circumstances, specific "population equivalents" calculated for particular industries per unit of raw materials or products, can only be used as general estimates. Joint treatment of industrial effluents with municipal sewage is also the subject of numerous

legal, financial and political questions. Depending on the type and characteristics of industrial effluents, some pretreatment is practised in the
industrial plant prior to the discharge of the effluents to publicly owned
treatment plants. The effectiveness of an industrial pretreatment program
to limit the discharge of toxic metals and organic substances was studied
recently by Ongerth and DeWalle (1980), and may be presented as an example of such procedures. Respective regulations formulate quality requirements for those discharges to protect both sewers and the treatment
plant itself. Completely restricted are flammable or explosive substances.
strongly toxic materials, solid wastes, and acids. Limited are discharges
of fats, greases and oils, heavy metals, taste- and odour-causing substances, high pH wastewater, and high BOD and suspended solids wastewater.

The U.S. Federal Guidelines (Section 307a of the 1977 Clean Water
Act) provide for the control of 65 specific pollutants. Various industrial
waste components presented as threshold concentrations in Table 8-1
and 8-2 have inhibitory effects on activated sludge treatment and the

TABLE 8-2

Inhibitory Threshold Concentrations of Some
Organic Components of Industrial Effluents
(in mg/L)

Pollutant	Activated Sludge Process	Nitrification Process
Allyl alcohol		19.5
Phenol	200(?)	4-10
Creosol		4-16
2-4, dinitro- phenol		150
Allyl Chloride		180
Dichlorophen		50
Thiourea		0.075
Thioacetamid		0.14
Trinitrotoluene	20-25	300
EDTA	25	
Pyridine		100
Benzidine	500	

nitrification process. These numerical values were determined as concentrations in influents to the treatment units and, therefore, could be open
to criticism.

The Green Bay Sewage Treatment Plant (Fig. 8-4) is an example of
joint wastewater treatment. Presently, this plant treats, in a contact
stabilization system, 22 mgd of sewage and 17.5 mgd of pulp and paper
mill effluents (83,270 and 66,238 cu.m per day, respectively). An ex-
tention of the plant is designed for an average flow of 52 mgd (196,820 cu.m
per day) and a peak flow of 171 mgd (647,235 cu.m per day). This case
study was presented by Barton et al. (1975). Another interesting example
of activated sludge joint treatment of various industrial effluents is the
Fukashiba Wastewater Treatment Plant for the Kashima Industrial Complex (Matsui et al., 1974). This plant accepts pre-treated effluents from
24 industrial plants and some sewage flow. Numerous components of the
industrial effluents are only partially biodegradable.

8.4 TREATMENT OF PHENOLIC WASTEWATER

Phenols are a very common component of many industrial effluents amenable to biological treatment. Phenol bearing wastewaters include: coke-
plant effluents, effluents from coal gasification plants, petrochemical
plants, and from the production of many synthetic organic compounds, etc.

Figure 8-4. Joint wastewater treatment plant in Green Bay, Wisconsin
(photo courtesy of Greey Lightnin).

In this work, removal of phenols from wastewater by means of the activated sludge process will be discussed, using coke-plant effluents as an example. Characteristics of these effluents were presented in p. 8.2.

At present, biological oxidation of phenols in coke-plant effluents is not considered a "difficult" technology. However, a very effective and consistent removal of phenols associated with a substantial decomposition of cyanide and thiocyanate requires relatively low substrate loadings and a long retention time in aeration tanks. Monomeric phenols undergo biological oxidation at a fairly rapid rate, while polymeric phenols require much longer aeration periods, because of their slow rate of biological oxidation. Moreover, the partially oxidized and highly polymerized phenolic compounds may be inhibitory to all oxidation reactions occurring in the activated sludge treatment of coke-plant effluents.

Most of the existing activated sludge plants for coke-plant effluents treatment were designed by selecting the mixed liquor aeration times and the mixed liquor suspended solids levels from the data available from pilot-plant experiments and from the operation of full-scale installations. A number of activated sludge plants for coke-plant effluents operate at elevated temperatures in the range of 25-30°C, and an important factor is maintaining the temperature of the mixed liquor as constant as possible. Moreover, the net yield of bacterial mass (excess activated sludge) in the treatment of coke-plant effluents is much smaller (per lb. of BOD removed) than respective values in the treatment of many other wastewater.

Some characteristic examples of activated sludge treatment of coke-plant effluents are presented as follows:

<u>DOFASCO Plant, Hamilton, Ontario (Fall 1975)</u>. The DOFASCO coke-plant produces an average of 3,680 t/day of metallurgical coke from a low-sulphur West Virginia coal. The biologically treated effluents (336,000 U.S. gpd = 1,272 cu.m per day) include excess crude flushing liquor, condensate from the primary cooler, condensate from final coolers, and condensate from the light oil plant. The wastewater pretreatment is based on at least 1.5 day storage and ammonia stripping with lime addition. Clarification after stripping is not satisfactory and substantial amounts of mineral suspensions are carried over to the aeration tanks for 23 hrs. aeration time and operating at about 30°C. No intentional removal of excess sludge is practised and the levels of mixed liquor suspended solids (MLSS) and mixed liquor volatile suspended solids (MLVSS) reach 23,000 and 8,000 g/cu. m (Fig. 8-5), respectively. This causes a solids overload on the final clarifier and, consequently, high suspended solids levels in the final effluent. Removals (Fig. 8-6) of phenols are excellent (in the range of 0.1 mg/L of residual phenol), but removals of thiocyanate are not effective enough (about 80 mg/L in effluent). No noticeable nitrification of ammonia occurs during this treatment (Ganczarczyk and Elion, 1978).

<u>U.S. Steel Plant, Clairton, Pennsylvania (Fall 1976)</u>. This plant
generates over 2.5 mgd (9,500 cu. m/d) of wastewater of which 45%
originates from the coking operation and 55% from the chemical processing of coking by-products. The treatment plant accepts wastewater after separate free- and combined-lime ammonia stripping and
50% dilution. Its basic design (by Arthur McKee & Co.) was for a
mixed liquor suspended solids level of 5,000 g/cu. m and 3 days aeration time. It was expected that, under such conditions, very advanced
nitrification of wastewater would occur. Because of numerous operational problems, in 1980 only up to 75% of nitrification was achieved.
Very light sludge was produced with limited flocculation abilities,
resulting in a suspended solids level in the final effluents of up to
200 mg/L. Therefore, filtration of the final effluent is being prac-

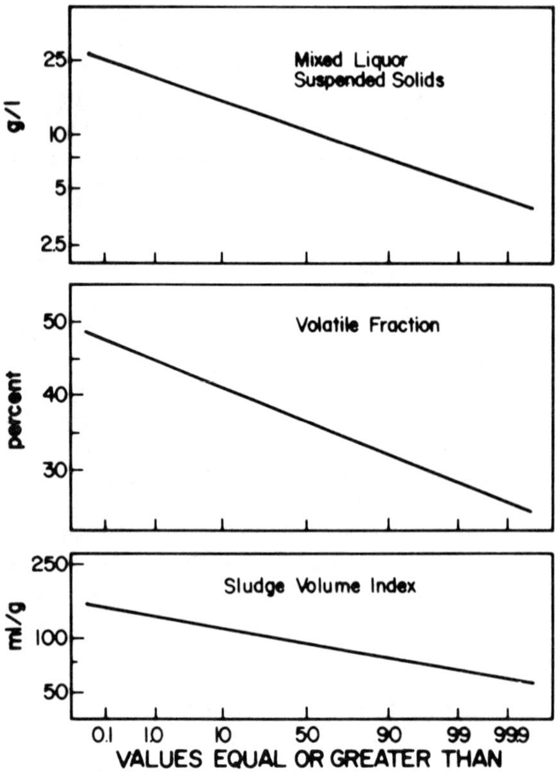

Figure 8-5. Distribution of treatment parameters data at DOFASCO coke-
plant effluent plant.

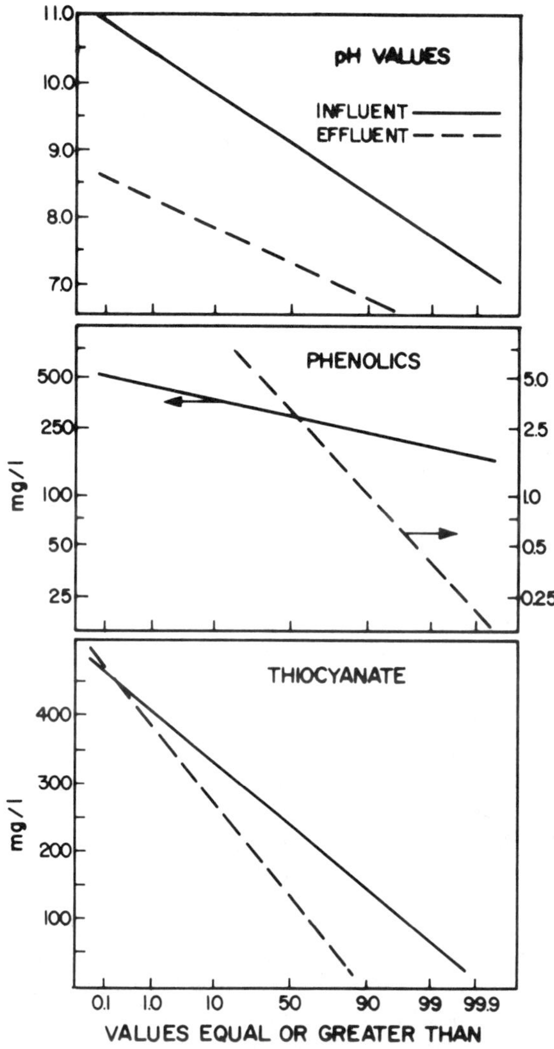

Figure 8-6. Distribution of treatment effects data at DOFASCO coke-plant effluent plant.

tised. The sludge wasting was designed for 5-10% of sludge inventory per day. The excess sludge is treated by the addition of ferric salts, lime and polyelectrolytes, and subsequent dewatering by vacuum filtration. The vacuum filter cake is disposed of by land filling. Some problems are also associated with this practice.

<u>Sumikin Coke Company Plant, Kashima, Japan (April 1977)</u>. The
plant is designed to treat 760,000 gpd (2,800 cu. m per day) of coke-
plant effluents, which originally contained 1,500-2,000 mg/L phenols.
For removal of residual oils and tar, air flotation is applied. The next
step in the treatment is a phenosolvan extraction of phenols, operating
with about 90% efficiency. Then, steam stripping of free ammonia is
applied, wastewater is diluted 2.5 times and treated by aeration with
activated sludge in 20 hrs. retention aeration tanks. The final effluents contain only trace concentrations of phenols (0.01 mg/L), about
3 mg/L of total cyanides, 1-2 mg/L of CNS, and about 1,000 mg/L of
residual ammonia. Nitrification in the aeration tanks occurs only
sporadically, producing maximum concentrations of 100 mg/L of
$N-NO_2$ (Kuwana, 1977).

<u>Bethlehem Steel Plant, Bethlehem, Pennsylvania (Summer 1977)</u>.
This plant accepts a maximum of 180,000 gpd (680 cu. m per day) of
coke-plant effluents which is first subjected to 5-7 days retention,
prior to a dilution 1:3 with water from heat exchangers. No ammonia
stripping from this wastewater is being practised. The activated
sludge treatment operates with a mixed liquor suspended solids level
of 2,500-4,000 mg/L, and with 35°C in the aeration tanks. The phenol
concentration in the final effluent is less than 0.5 mg/L, but joint concentrations of total cyanide and thiocyanate reach 15 and 250 mg/L,
respectively.

<u>Armco Hamilton Works, New Miami, Ohio</u>. This plant was put into
operation only in 1980 (Wear et al., 1980), and it is treating waste
ammonia flushing liquor and benzol plant effluents diluted by small
flows of sewage and river water (total dilution 1:1). Waste ammonia
liquor is stripped off ammonia with the use of caustic soda for alkalization, and benzol plant wastewater is subjected to a gravity oil
separation. Activated sludge treatment is performed in aeration
tanks with integral clarifiers, and at a minimum aeration time of
24 hrs. The mixed liquor suspended solids are in the range of
15,000 mg/L. The effluent residual phenol concentrations are as low
as 0.001 mg/L. The excess of sludge is directed to the hot gases exhaust and then incorporated into tar which is used as a fuel.

8.5 TREATMENT OF PULP MILLS EFFLUENTS

Activated sludge treatment of pulp mill effluents represents a substantial
percentage of the processes used for the treatment of wastewater other
than municipal sewage. Most of the biological treatment plants for pulp
mills are for wastewater from Kraft pulp mills and semichemical pulp
mills. Recently, sulfite pulp mill effluents have also been subjected to

this treatment because of the chemicals recovery made possible by the use
of soluble bases pulping.

Activated sludge treatment easily removes biodegradable organic substances from pulp mill effluents, but has a limited effect on COD, colour
and toxicity removals. The mechanism of lignin removal during activated
sludge treatment of pulp mill effluents (Ganczarczyk and Obiaga, 1974) is
a complex phenomenon, which covers physical sorption, perhaps chemical
oxidation and condensation, and some biological degradation. In the treatment of simulated sulfite pulp mill effluents, lignin was removed primarily
as lower molecular fractions of lignosulfonates. A synthesis of higher
molecular weight lignin material seemed possible, while during the treatment of Kraft mill effluents, lignin was removed almost uniformly throughout the whole spectrum of molecular weight distribution (Fig. 8-7). In all
the treatment cases of both types of effluents which were studied, the respective activated sludges contained substantial amounts of retained lignins.
Alkali-lignin were more effectively removed than lignosulfonates. This

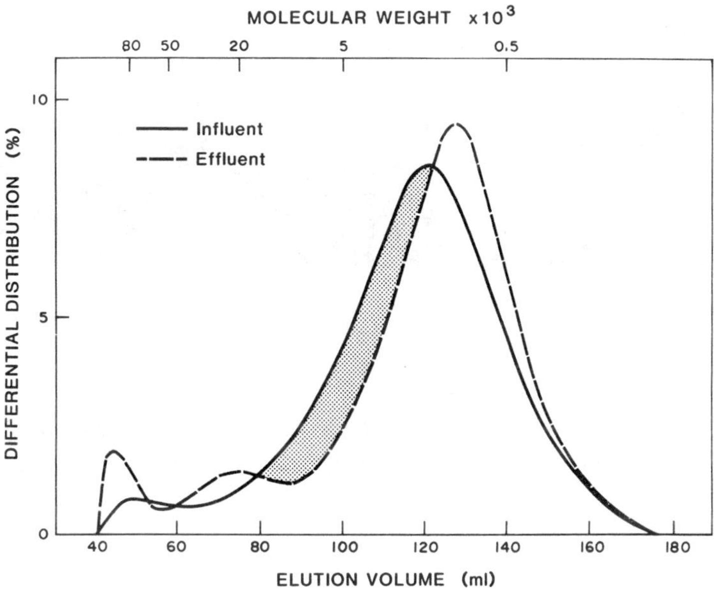

Figure 8-7. Typical changes in molecular weight distribution of alkali-
lignin in activated sludge treatment of simulated Kraft pulp mill wastewater (M.W. = molecular weight).

Figure 8-8.   Treatment plant for Kraft pulp mill effluents in Covington, Virginia (photo courtesy of Westvaco).

may be related to the larger amount of the hydroxyl group present in the alkali-lignin macromolecules and the hydrophilic properties of the sulfonic group of lignosulfonates. The maximal values observed for alkali-lignin removals were 70 percent, while lignosulfonates removal reached only 15.2 percent.

A number of resin acid derivatives and chlorinated lignin breakdown products present in these effluents have acute toxic effects on fish. It appears that the compounds responsible for most of this toxicity are mono- and dichloro-dehydrobietic acid, tri- and tetrachloro-guaiacol and epoxy-steric acid.

The current state of the activated sludge process technology for pulp mill effluents treatment is shown in the following examples:

<u>Westvaco Plant in Covington, Virginia</u>. This plant (Fig. 8-8) was
built in 1955 for the treatment of effluent from the production of
bleached Kraft and semi-chemical pulp; it was the first, full-scale
activated sludge installation operating without any admixture of sewage
but with the addition of mineral nutrients instead (Pearman and Burns,
1957). The original plant was designed for a wastewater flow of
16 mgd (60,560 cu. m/d) and its current capacity, after several additions and improvements, is 26 mgd (94,625 cu. m/d)

The treatment sequence is as follows (Fig. 8-9): The all wastewater streams (including those from paper and board making operations
and activated carbon production), as well as the excess activated sludge
and the filtrate from vacuum filtration of sludge, are first collected in
a mixing tank for a minimal retention of 2.5 min. and neutralized with
lime or sulphuric acid. Then, the removal of suspended solids from
the mixed influent follows. Two primary clarifiers, each 100 feet
(30.5 m) in diameter, are equipped with Dorr-Oliver scrapers. At
the 25 mgd (94,625 cu. m/d) flow rate the wastewater detention time

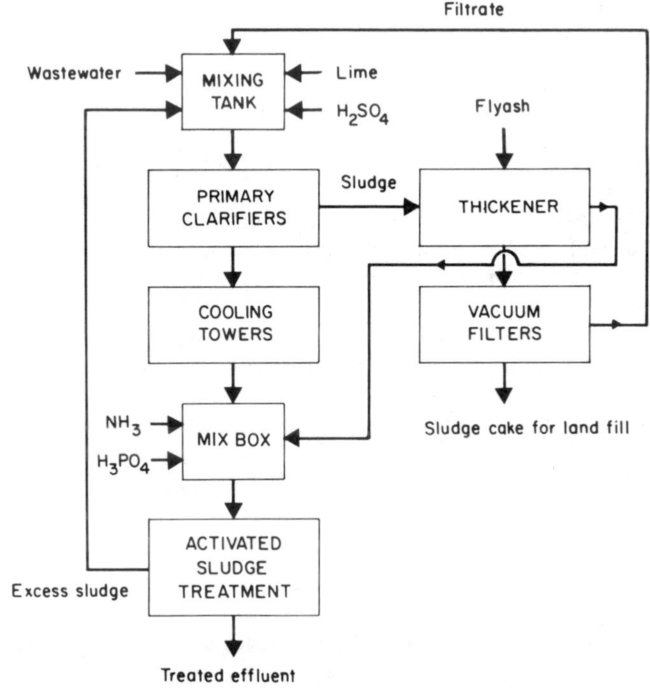

Figure 8-9. Flow-diagram of the Covington treatment plant.

in the clarifiers is 1.6 hrs. and the surface settling rate is about
1,600 gpd/sq. ft. (65 cu. m/d/sq. m). The primary sludge is pumped
to a thickener, and the primary effluent is conveyed to cooling towers
if the wastewater temperature is in excess of 105°F (40.6°C). At this
plant, it was established (see p. 3.7) that an optimum operation of
activated sludge treatment required a mixed liquor temperature of
about 100°F (38°C). However, the summer temperatures of the primary clarifier effluent can reach 120°F (49°C). The cooling system
in the Covington plant is composed of three forced draft cells, each
employing an adjustable pitch, propeller-type fan of 18 feet in diameter and operating at 177 rpm. The wastewater is pumped to the
top of the cooling towers and flows by gravity to the mix box, and then
to the two aeration tanks, which are each composed of two aeration
bays, and subsequently to the two secondary clarifiers. In the mix
box, the wastewater flow is joined by the supernatant from the primary
sludge thickener, and enriched with mineral nutrients: ammonia and
phosphoric acid. The nutrient addition is carried out to achieve BOD:
N:P ratio equal to about 100:8:0.7. The total volume of aeration tanks,
designed as aeration bays around secondary clarifiers is 2.8 mg
(10,600 cu. m). The aeration system applied is composed of coarse
bubble sparger rings located beneath double disc turbines (submerged
and surface). Each aeration bay is equipped with 5 sparger turbines
(the first three are 25 HP and the final two are 20 HP). The total capacity of the compressed air system is 14,600 SCFM (413N cu. m/min).
The total volume of the two secondary clarifiers is 1.15 mg (4,353 cu.
m). Each is of the peripherial feed type, and 120 feet (36.6 m) in
diameter.

The characteristic operation data for wastewater treatment at this
plant in March 1976 is as presented in Table 8-3 (Burns, 1976). The
BOD removal is almost 89% and the COD removal is over 75%. Although the suspended solids removal is almost 90%, the residual level
of suspended solids of 120 mg/L shows somewhat limited flocculation
ability of the activated sludge micro-organisms, which is apparently
characteristic of the type of wastewater treated.

The sludge treatment at the Covington Plant is accomplished by
directing the excess activated sludge to the mixing tanks, gravity
thickening the primary sludge with flyash, and dewatering the thickened
sludge in the vacuum filters. The respective flows and solids concentrations are presented in Table 8-4.

Ostroleka Plant. The activated sludge treatment plant in Ostroleka,
Poland, was designed for the treatment of 4.7 mgd (17,800 cu. m/day)
of unbleached Kraft pulp mill effluents with a small admixture of
primary treated municipal sewage. This plant was put into operation
in 1965, and during the operation in 1968, it received 2.4-3.2 mgd
(9,000-12,000 cu. m/day) of mill effluents, and 0.26-0.53 mgd (1,000-

TABLE 8-3
Operation of the Covington Plant
Wastewater Treatment

Influent:

Flow	21.1	mgd (79,864 cu. m/d)
Suspended solids	1,150	mg/L
BOD	420	mg/L
COD	1,610	mg/L
Temperature	108	°F (42.2°C)

Primary Effluent:

Suspended solids	210	mg/L
BOD	280	mg/L
BOD filtered	210	mg/L

Aeration Tanks:

MLSS		1,700	mg/L
SVI		59	mg/L
Sludge load		1.4	gBOD/gMLSS/d
Volumetric load		146	lbBOD/1,000 cu. ft/d (2,336 gBOD/cu. m/d)
Solids production		0.63	g/gBOD rem.
Oxygen Uptake Rate:	in	63	mgO_2/gMLSS/hr
	out	11	mgO_2/gMLSS/hr

Secondary Effluent:

Suspended solids	120	mg/L
BOD	47	mg/L
BOD filtered	15	mg/L
COD	400	mg/L

2,000 cu. m/day) of municipal sewage (Ganczarczyk, 1969). The pulp mill effluents were first collected in an equalization pond of a volume of 3.7 mg (14,000 cu. m) and then conveyed through a 3.1 mile (5 km) long pipeline to the treatment plant, where they were mixed with mechanically treated sewage and enriched with mineral nutrients to the ratio BOD: N:P equal to 100:4:0.7. The treatment plant incorporates six parallel, double, low-pressure aeration tanks (modification of the Inka system) and 4 clarifiers. Each aeration tank has a volume of 177,000 gal (670 cu. m).

In order to obtain more detailed information about this plant's performance, 99 series of performance measurements were taken in the summer of 1967 and again in the summer of 1968 (Ganczarczyk,

TABLE 8-4
Operation of the Covington Plant
Sludge Treatment

	FLOW		SOLIDS
	mgd	cu.m/d	
Primary underflow	0.9	3,400	1.7%
Thickener underflow	0.82	3,100	2.6%
Thickener overflow	0.71	2,700	540 mg/L
Filtrate	0.62	2,350	740 mg/L
Sludge cake	--	---	23.4%

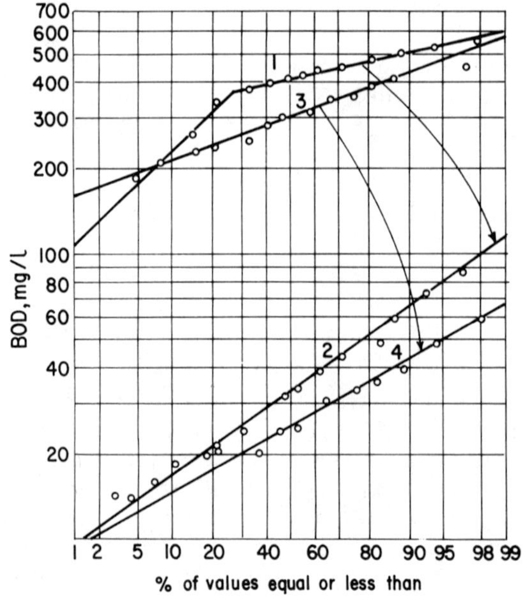

Figure 8-10. Distribution of BOD data in the Ostroleka Plant (1—untreated
wastewater in Period I, 2—treated wastewater in Period I, 3—untreated
wastewater in Period II, 4—treated wastewater in Period II).

1969). Figures 8-10 and 8-11 present the observed BOD and COD removals, respectively. Average performance data for both study periods are given in Table 8-5.

<u>OJI Plant in Tomakomai, Japan</u>. The effluents at the Oji Pulp and Paper Plant in Tomakomai, Hokkaido (Japan) are composed of magnesium-sulphite and semi-chemical pulping wastewater, condensates from chemicals recovery and bleaching wastewater. They are originally slightly acidic (pH around 6) and are at first neutralized to pH = 8.5 by addition of caustic soda. The average BOD of these effluents is about 500 mg/L and their COD is about 1,000 mg/L. A relatively constant year round wastewater temperature of 33-35°C is controlled by a cooling tower. The treatment of the effluents is performed by means of the oxygen activated sludge process (UNOX system), with a retention time of 2.6 hours at a mixed liquor suspended solids level of 5,000 mg/L. During the author's visit to this plant (April 1977), the dissolved oxygen in the aeration tanks was maintained at a very high level of 10-15 mg/L because the capacity of oxygen generation unit exceeded current needs. The treatment plant ef-

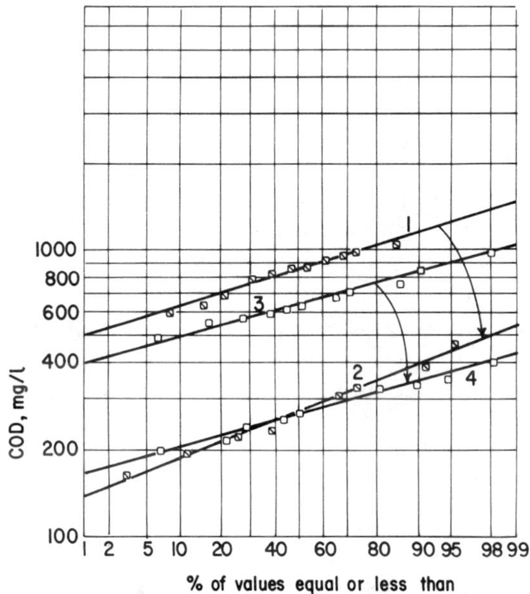

Figure 8-11. Distribution of COD (simplified) data in the Ostroleka Plant (1, 2, 3 & 4 - as in Fig. 8-10).

TABLE 8-5
Operation of the Ostroleka Plant

	Period I	Period II
Volumetric loading		
lb BOD/1,000 cu. ft/day	89	64
g BOD/cu. m/day	1,430	1,030
Sludge loading		
g BOD/gMLSS/day	0.42	0.26
Sludge age, days	10.4	7.1
BOD removal, percent	90.5	90.9
COD (simpl.) removal, percent	66.5	58.6

fluent BOD was in the range of 30-40 mg/L, the COD about 400 mg/L, and the residual suspended solids level was around 40 mg/L. The excess sludge was about 0.6 kg/kg BOD removed. Operational experience at this plant showed that the system removed a relatively constant percentage of the influent BOD but responded to the fluctuations of the influent concentrations. It was also observed that the high dissolved oxygen levels in the aeration tanks helped to maintain good treatment effects. A decrease of the dissolved oxygen level to 5-6 mg/L (as suggested by Union Carbide) caused some small decrease in the quality of the plant effluent. A comparison between the Oji plant operation and the Westvaco (Covington, Virginia) plant is presented in Table 8-6.

KAO Plant in Fuji, Japan. At the Kao-Fuji Pulp and Paper Plant in Fuji, Shizuoka Prefecture, effluents from semi-chemical pulping, de-inking, and condensates from chemicals recovery have an average BOD of about 350 mg/L and temperature of 30-32°C. The treatment by means of oxygen activated sludge (UNOX system) is characterized by an average retention time in aeration tanks of 2.5 hours at a mixed liquor suspended solids level of about 8,500 mg/L. The effluent BOD is usually somewhat over 10 mg/L with occasional increases to 30 mg/L. The residual suspended solids level in final effluent fluctuates up to 60 mg/L. A comparison of this plant operation with that of the Ostroleka plant is presented in Table 8-7.

TABLE 8-6

Comparison of Oji-Tomakomai and
Westvaco-Covington Plant Operations

Plant	OJI in Tomakomai	WESTVACO in Covington
Treatment Parameters :		
F/M	0.92	1.4
MLSS, g/cu.m	5,000	1,700
Vol. loading, gBOD/cu.m/d	4,615	2,336
Retention time, hrs.	2.6	2.7
Temperature, °C	33-35	below 38
Dissolved oxygen, mg/L	10-15	up to 2
Effluent Quality:		
BOD, mg/L	500 → 30-40	280 → 47
COD, mg/L	1,000 → 400	→ 400
Residual SS, mg/L	40	120
Sludge Production:		
kg/kg BOD removed	0.6	0.6

8.6 OTHER EXAMPLES OF INDUSTRIAL
WASTEWATER TREATMENT

There are numerous other examples of activated sludge treatment of
various industrial effluents, and it is very difficult to arrange these examples in a logical way. Some of the examples are very unique and some
show similarities with the activated sludge plants for phenolic wastewater
or pulp mill effluents treatment which were discussed previously. A few
arbitrarily selected examples of the former type of activated sludge treatment are as follows:

TREATMENT OF FLAX RETTING WASTEWATER IN AN
OXIDATION DITCH

The anaerobic retting of flax causes the development of a dark brown liquor
which has a characteristic strong smell, low pH values, and contains

TABLE 8-7
Comparison of Kao-Fuji and
Ostrolenka Plant Operations

Plant	KAO in Fuji	OSTROLENKA
Treatment Parameters:		
F/M	0.4	0.4
MLSS, g/cu.m	8,500	3,300
Vol. loading, g/BOD/cu.m/d	3,360	1,430
Retention time, hrs.	2.5	6.8
Temperature, °C	30-32	17-22
Dissolved oxygen, mg/L	5-6	0-4
Effluent Quality:		
BOD, mg/L	350 → 10-30	400 → 38
Residual SS, mg/L	up to 60	4-100

considerable quantities of organic acids. For the treatment of such liquor
in the Mieroszow Mill (Poland), an oxidation ditch has been used successfully. The treatment plant consists of a strainer, the oxidation ditch, and
a secondary clarifier. The oxidation ditch has a volume of 68,700 gal
(260 cu.m), a liquid depth of 3 ft (0.9 m), and is equipped with four
Kessener brushes with a length of 16.4 ft (5 m), coupled with motors of
10 kW each. This ditch is loaded once per day with 15,825 gal (60 cu.m)
of flax retting wastewater which flows to the unit after the periodic stop-
page of the work of the brushes and the decantation of 7.9 in (20 cm) level
of treated effluent. Normally, two Kessener brushes operate in the ditch,
with the other two used as a reserve. As a result of this treatment, flax
retting wastewater BOD decreases from 1,000-3,000 mg/L to 25-60 mg/L.

LURGI-MITSUBISHI ACTIVATED SLUDGE PROCESS

The Mitsubishi Oil Company petroleum refinery located in the centre of the
Mizushima Industrial area (Japan), has the capacity to process 270,000
barrels of crude oil per day. The objectives for the quality of the treated
wastewater are based on a required COD in the receiving water body of
2-3 mg/L and a non-dectable n-hexan extractable substance (soluble oils).
The characteristics of the wastewater show COD in the range of 87-
350 mg/L, with an average concentration of 120 mg/L, and oil content in

the range of 15-250 mg/L, with an average concentration of 150 mg/L.
The treatment of this wastewater should produce effluent with an average
COD of 30 mg/L (max. 40 mg/L) and oil contents of 1.0 mg/L (max.
2.0 mg/L). The wastewater is treated first in an API Oil Separator and
then is collected in an equalization tank of 5,000 cu. m volume (at least
1 day retention). The next treatment step is coagulation with $FeCl_3$ (to
break up oil emulsion) and removal of flocs by air flotation. To control
pH of flocculation at 7.0, a caustic soda solution is added. About 40% of
the effluent from flotation is pressurized with air to 3 kg/cm^2. The suggested surface load of the flotation unit is 4.4 cu. m per sq. m per hr.
The average consumption of 38% $FeCl_3$ is 632 kg/day. The overflow from
the flotation unit is directed to the activated sludge treatment unit designed
for the load of 480 g BOD per cu. m of aeration tank per day. The intended mixed liquor suspended solids level is about 5,000 ppm. and the
secondary clarifier operates at an hydraulic load of 0.59 cu. m/sq. m/day.
This activated sludge system operates in accordance with the Lurgi-
Mitsubishi system which calls for an addition of methanol to the wastewater
stream to maintain the influent BOD level at a relatively constant value of
100 ppm. In this way, a continuous production of bio-mass is secured and
perhaps, a stimulation of biological oxidation of soluble oils occurs. The
average addition of methanol is about 290 kg/day. Moreover, a phosphorus
nutrient is added as 20.3 kg/day of 75% H_3PO_4. The outflow from the
secondary clarifier of the activated sludge treatment process is subjected
to mechanical flocculation in a "Hi-Aqua-Finer" installation to convert the
activated sludge suspended solids, which are carried over, into a settle-
able form. The last step in the wastewater treatment is a holding tank
with a capacity for several days retention of the wastewater flow. This
tank also receives the surface runoff from the terrain of the refinery.

PHARMACEUTICAL WASTEWATER TREATMENT PLANT
IN FUKUROI, JAPAN

The Nippon-Roche Company's Fukuroi Plant produces mostly vitamin B_2.
The 750-1,000 cu. m per day of the plant wastewater (year 1977) was first
chemically neutralized (from pH of about 4 to pH of 5.8), then clarified,
subjected to activated sludge treatment, and eventually treated by granu-
lated activated carbon filtration. Due to this treatment, the initial wastewater BOD of 4,500 mg/L decreased to 10 mg/L, its permanganate COD
of about 3,000 mg/L was reduced to 30 mg/L, and its strong brown colour
was completely removed. The applied wastewater aeration time in the
activated sludge process was in the range of over 200 hrs. (8.3 days),
and the maintained mixed liquor suspended solids level was 3,000 mg/L
in winter and 800 mg/L in summer. The resulting sludge retention time
(sludge age) was about 10 days, and the sludge volume index was around
100 mL/g. The specific problems faced in operation of this plant was poor
sludge dewatering and air diffusors clogging requiring chemical treatment

every few weeks. The cost of such extensive wastewater treatment in 1977
was reaching up to 8% of the plant production value, but it was expected
that this factor would improve with the expansion of the plant production
(Mukai, 1977).

REFERENCES

Barton, C. A. et al.: "Joint Treatment of Pulping and Municipal Wastes,"
Jour. Water Poll. Control Fed., 47, 998 (1975).

Brown, M. J., and Lester, J. N.: "Metal Removal in Activated Sludge:
The Role of Bacterial Extracellular Polymers," Water Research (Brit.).
13, 817 (1979).

Burns, O. B., Jr.: personal information (1976).

Eckenfelder, W. W., Jr.: "Theory of Biological Treatment of Trade
Wastes," Jour. Water Poll. Control Fed., 39, 240 (1967).

Environmental Protection Agency: "Federal Guidelines, (Draft), State
and Local Pretreatment Programs," Vol. I, Contract No. 68-01-2963,
(1975).

Ganczarczyk, J.: "Performance Studies of the Unbleached Kraft Pulp
Mill Effluent Treatment Plant in Ostroleka," Water Research (Brit.).
3, 519 (1969).

Ganczarczyk, J., et al.: "Depolymerization of Lignosulfonic Acids in
Sulfite Pulp Mill Effluents," paper presented at the 47th Annual Conference
of Water Pollution Control Federation, Denver, Colorado, 1974.

Ganczarczyk, J., and Obiaga, T.: "Mechanism of Lignin Removal in
Activated Sludge Treatment of Pulp Mill Effluents," Water Research
(Brit.), 8, 857 (1974).

Ganczarczyk, J., and Elion, D.: "Extended Aeration of Coke-Plant Effluents," Proc. 33rd Ind. Waste Conference, Purdue University, 895
(1978).

Ganczarczyk, J.: "Pre-treatment of Coke-plant Effluents," in "Waste
Treatment and Utilization," pp. 119-126, Pergamon Press, Oxford, 1980.

Kuwana, Yasuo: personal information (1977).

Ludzack, F. J. and Ettinger, M. B.: "Estimating Biodegradability and
Treatability of Organic Water Pollutants," Biotechnol. Bioeng., 5, 309
(1963).

Matsui, S. et al.: "Activated Sludge Degradability of Organic Substances
in the Wastewater of the Kashima Petroleum and Petrochemical Industrial

Complex in Japan," paper presented at the 7th Inter. Conference on Water Pollution Research, Paris 1974.

Mukai, T.: personal communication (1977).

Mytelka, A. I. et al.: "Heavy Metals in Wastewater and Treatment Plants Effluents," Jour. Water Poll. Control Fed., 45, 1859 (1973).

Narkis, N., and Schneider-Rotel, M.: "Ozone-Induced Biodegradability of a Non-Ionic Surfactant," Water Research (Brit.), 14, 1225 (1980).

Oliver, B. G., and Cosgrove, E. G.: "The Efficiency of Heavy Metal Removal by a Conventional Activated Sludge Treatment Plant," Water Research, 8, 869 (1974).

Ongerth, J. E., and DeWalle, F. B.: "Pretreatment of Industrial Discharges to Publicly Owned Treatment Works," Jour. Water Poll. Control Fed., 52, 2246 (1980).

Pearman, B. V., and Burns, O. B.: "Activated Sludge Treatment of Kraft and Neutral Sulphite Mill Wastes," Sew. Ind. Wastes, 29, 1145 (1957).

Pitter, P.: "Determination of Biological Degradability of Organic Substances," Water Research, 10, 231 (1976).

Sterritt, R. M., and Lester, J. N.: "The Influence of Sludge Age on Heavy Metal Removal in the Activated Sludge Process," Water Research (Brit.), 15, 59 (1981).

Tokuz, R. Y., and Echenfelder, W. W., Jr.: "The Effect of Inorganic Salts on the Activated Sludge Process Performance," Water Research (Brit.), 13, 99 (1979).

Volesky, B. et al.: "The Effect of Metallic and Ionic Species on the Performance of a Biological Effluent Treatment System," Proc. 12th Canad. Symp. on Water Poll. Res., 12, 191 (1977).

Water Pollution Control Federation, Technical Practice Committee: "Joint Treatment of Industrial and Municipal Wastewaters," Washington, D.C., 1976.

Wear, M. R., et al.: "Biological Treatment of Coke-Plant Waste Utilizing an Integral Clarification Concept," Proc. 35th Industrial Waste Conf., Purdue University, 343 (1980).

Process Design

9.1 GENERAL

The technological design of an activated sludge process can be based on experience in the application of the process (see Chapters III and IV), and/ or the more specific information derived from laboratory or pilot-scale experiments carried out on the wastewater in question. Some generalization of this accumulated knowledge and experience may be presented in the form of many different mathematical models, ranging from simple process kinetics expressions to more sophisticated economical optimization formulas.

The activated sludge models, generally, may be divided into "deterministic" and "probabilistic" types. Deterministic models are those in which process variables can be presented as definite numbers for any given set of conditions. For the other models, the statistical techniques have to be used to express the principle of uncertainty. It appears that the real process has stochastic features. Therefore both approaches are valid and, for practical applications, should be combined or superimposed.

The most common approach to modelling of the activated sludge process is based on the application of the continuous-culture theory. The general Monod equation (Monod, 1949) is often applied to modelling the activated sludge process. This equation may be modified to account for the nonbiodegradable solids present in the process influent and/or formed during the process, but both its theoretical and practical meaning is quite limited.

Dynamic mathematical modelling of the activated sludge process is the best available technique for predicting the process outputs, i.e., the quality of the effluent. Although substantial progress was made recently in this field, there are still major deficiencies in the available models, which prevent the satisfactory prediction of the concentration of the suspended solids in the effluent from the secondary clarifiers and the resulting BOD. This

229

is mostly due to somewhat unclear relationships between the flocculation
ability of activated sludge and the process parameters.

9.2 BASIC DESIGN APPROACH

The early process design for activated sludge treatment was based on aeration detention time. Organic loading per unit of aeration volume was later
developed as a more basic approach. These two design approaches were
purely empirical. The first fundamental approach applied was in estab-
lishing food to micro-organism ratio as a design parameter (see Chapter III).

It is usually a requirement of a process design to determine substrate
removal, solids accumulation and oxygen requirement. In 1954, Eckenfelder and O'Connor proposed simplified mathematical relationships for
activated sludge treatment of wastewater. This approach was subsequently
modified and expanded by Eckenfelder (1961, 1966, 1967, 1970), and is now
widely accepted and widely used. Details of this approach were discussed
in Chapter II. Different formulations for the same relationships were offered by McKinney (1962). Their application was studied later by Burkhead
and McKinney (1968). In 1974, Goodman and Englade compared the latter
approach with the Eckenfelder's process design and concluded that both
methods could be translated to each other.

A broad incorporation of the sludge age (or mean cell residence time)
concept into the design of the activated sludge process was suggested by
Lawrence and McCarty (1970), Sherrard and Lawrence (1973), Stensel and
Shell (1974) and others. Although this approach has its merits, it cannot
be used as an exclusive guideline for many applications.

Process design in the field of activated sludge treatment is not yet an
exact science. Therefore, engineering experience and judgement are
probably the most important requirements for a successful approach to the
subject.

9.3 PROCESS ECONOMICS

The total costs for activated sludge treatment of wastewater are composed
of capital costs (such as the costs of land acquisition, construction of facilities, equipment, instrumentation, engineering, financing, etc.), and operational costs (including electrical power, labour, chemicals, etc.).
Specific cost analyses also take into consideration available grants, amor-
tization costs and other direct and indirect costs, in order to conclude the
calculation in the form of "equivalent annual costs." The concept of "value
engineering" is based on a life costs analysis of particular facilities, and
on updating the cost estimates at several points during the design stage.
For the correct design of a treatment plant, these costs should be

TABLE 9-1
Operating Costs Comparison for Brewery Wastewater Treatment
(after LeClair, 1982)

Components	In Canadian $ per kg BOD Removed	
	Deep-Shaft	Extended Aeration
Urea	0.045	0.025
Diammonium Phosphate	0.028	0.006
Polymer	0.034	–
Sulphuric Acid	0.006	–
Labor	0.100	0.062
Power	0.024	0.043
Maintenance	0.022	0.018
Sludge Hauling and Disposal	0.105	0.092
Total	0.364	0.246

minimized, without sacrificing the intended effects or the reliability of the treatment. The least costly design is usually achieved by the application of various technological trade-offs which are the most suitable for the specific situation. Very seldom will a particular technological or engineering solution prove cost effective for a wide range of needs. An example of the operating costs comparison for a brewery wastewater treatment by the deep-shaft technology and extended aeration process is shown in Table 9-1 (LeClair, 1982). Although power costs for the deep-shaft treatment are only about one-half of those for the extended aeration treatment, the latter process is less expensive due to the savings on nutrient requirements and on the labor.

Cost functions for different modifications of the activated sludge process and their elements were presented and analyzed by several authors. The Environmental Protection Agency (1978a) published capital costs figures in mid-1977 dollars for activated sludge process aeration tanks in municipal plants constructed from 1973-1977 (Fig. 9-1). Similarly, operation and maintenance costs for activated sludge treatment are presented in Figure 9-2 (Environmental Protection Agency, 1978b). The historical development of the cost function approach to treatment cost estimation for the preliminary design of treatment plants was discussed by Tihansky (1974).

Parkin and Dague (1972) developed a mathematical model for optimization of the whole system of wastewater treatment comprising primary

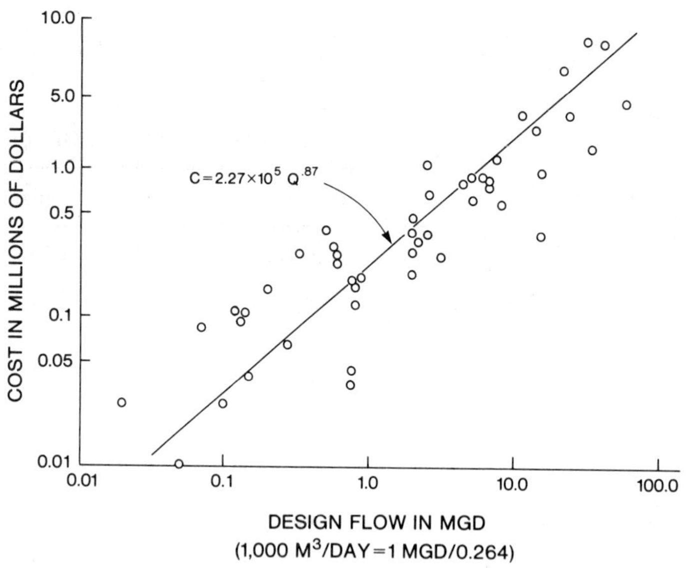

Figure 9-1. Construction cost function (in 1977 U.S. $) for activated sludge aeration tanks (after Environmental Protection Agency, 1978a).

sedimentation, the activated sludge process (according to McKinney's equations) and anaerobic or aerobic sludge stabilization. Their calculations showed that the effectiveness of primary clarification was of secondary importance; optimal mixed liquor suspended solids levels were at about 3,000 mg/L; and the costs of aerobic and anaerobic sludge stabilization were comparable. Many other similar studies were presented by several authors (CIRIA, 1973, Middleton and Lawrence, 1976, etc.). For European conditions, Floegl (1980) analyzed cost figures for eight different modifications of wastewater treatment, by means of activated sludge process, in plants servicing the equivalent of 10,000 to 500,000 people. He found that the required treatment efficiencies had only a secondary influence on the treatment costs. For plants below the equivalent of a population of 10,000, aerobic stabilization of sludge was practical, and for plants over the equivalent of a population of 25,000, primary clarification and sludge loading of 0.15 g BOD/g MLSS d., were advantageous. Generally, the wastewater flows had a primary influence on the treatment costs, with wastewater suspended solids levels and the nitrogen contents of wastewater proving less important. Cost effective tanks and other structures were constructed in a compact way, with a small length to width ratio, and possibly small depth below the ground water level. The cost-optimal depths of the tanks were found to be between 6 and 12 ft. (2 and 4 m), and, for circular tanks, ratios between diameter and depth in the range from 10:1 to 40:1 were most cost-effective. The comparison of aeration costs by compressed air or by surface aeration, did not show any decisive preference.

9.4 COMPUTERIZED DESIGN

As early as 1968, Smith published a program for the preliminary design of conventional wastewater treatment systems. Later in the same year, this program was expanded (Smith et al., 1968). The relationships used for activated sludge in these programs were for conventional modification of the process. A more advanced program was developed by Smith (1970) for a completely mixed activated sludge process. This work eventually led to the development of EXECUTIVE, a computer program which simulated performance and evaluated the cost of a number of wastewater and sludge treatment possibilities. Subroutines from the above program were also

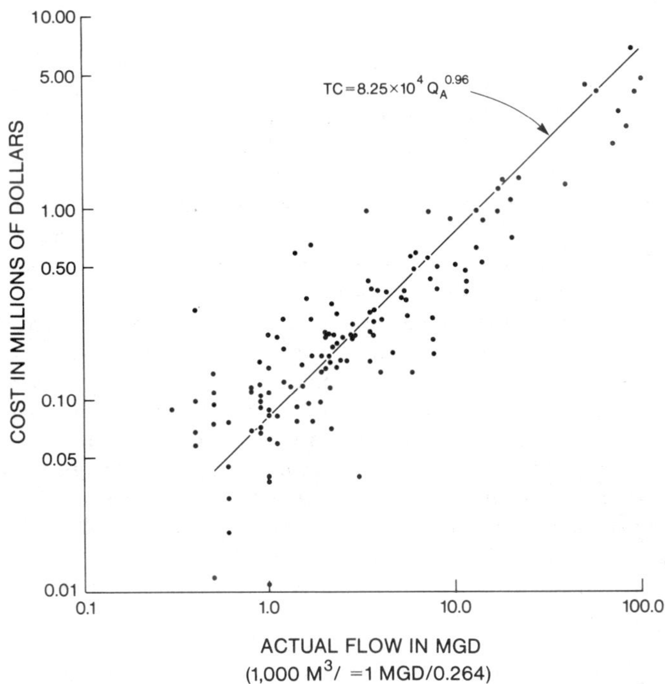

Figure 9-2. Operation and maintenance cost function (in 1977 U.S. $) for activated sludge process facilities (after Environmental Protection Agency, 1978b).

used for a computer-aided design procedure for the preliminary synthesis
of a wastewater treatment and sludge disposal system (Rossman, 1979).

CAPDET (Green et al., 1979) is another computer-based design procedure which can be used to select viable wastewater treatment trains, including the activated sludge process, to meet given effluent criteria, and
to rank the selected trains according to the lowest annual costs.

9.5 PROCESS ENGINEERING

The conversion of the results of the technological design of an activated
sludge treatment system into an engineering design of an appropriate treatment plant requires a number of engineering decisions, based on regulations, experience and economics. Various design models and programs
are basically only good for initial planning purposes. The engineering design should be always done by a skilled and experienced consulting engineer,
in accordance with applicable design parameters and/or pilot study results.
<u>Selection of the design modification</u>. Treatment requirements and treat-
ment economics are the basis for selection of the most suitable process
modification (see Chapter IV).
<u>Design Data</u>. Selection of design parameters for activated sludge treatment
requires considerable engineering experience. Treatment units designed
at high parameters are less flexible for fluctuating system loadings, and
are more prone to operational upsets. On the other hand, the selection of
treatment parameters that are too low, makes the treatment plant unnecessarily expensive, and may also result in some deterioration in the treatment results.
<u>Plant flexibility</u>. Perhaps, the most important factor to be considered in
the design of activated sludge plants is the desired degree of plant flexibility.
This, in turn, influences the number of aeration tanks and secondary
clarifiers (treatment units), the selection of the aeration tank flow pattern,
aeration system, etc. There is no simple rule for determining the optimal
number of treatment units, but, except for small plants, it is often suggested that at least three or four separate units be designed. If one unit
is out of service, the remaining 67 or 75 percent of the design capacity is
still available. Designing only two treatment units, may sometimes lead
to the over-dimensioning of their sizes and equipment. Multi-unit design
allows for the operation of only a part of the plant when the load being received calls for it. On the other hand, however, increased capital and
operational costs can arise from plant compartmentalization which is too
excessive. Sizing of the treatment units is closely related to the selected
number of units. However, some dimensions of the units are more
privileged than others. For instance, the depth of aeration tanks usually
does not surpass 15 ft. (3 m). Among other things, this is dictated by the
characteristics of the available blowers. When deeper tanks are decided

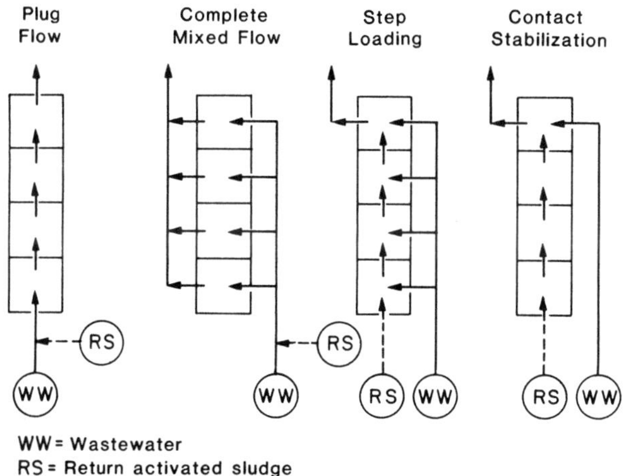

Figure 9-3. Aeration tanks layouts for different flow patters of mixed liquor.

upon, for example because of scarcity of land, an individual analysis is required for such a project.

The flexibility of an activated sludge plant can be increased by designing aeration tanks in such a way as to make possible some operational changes in the mixed liquor flow patterns (Laughton, 1980). Figure 9-3 presents an example of aeration tanks layouts for plug-flow, complete mixing, step loading (also called "step aeration") and contact stabilization. Configuration of the Plant. The treatment units which make up the activated sludge treatment plant can be arranged differently at the plant location. The idea of "blocking" or "nesting" of the facilities by the use of common walls helps to decrease construction costs, land acquisition requirements, and heat losses from wastewater during treatment. The latter feature may be very important in countries where there are severe climatic conditions. Figure 9-4 shows an example of an activated sludge treatment plant with some facilities blocked into one concrete structure. The selected configuration of facilities should allow for future plant expansion and/or modification. This plant also has the operational flexibility noted in Figure 9-3. Climatic Requirements. Cold climate conditions affect wastewater activated sludge treatment in several ways, including a decrease in sludge settleability, some inhibition of nitrification, etc. Above all, the freezing phenomena in secondary clarifiers are detrimental to the treatment results. Therefore, designs for some of these plants include covers on the secondary clarifiers (Figs. 9-5 and 9-6).

Figure 9-4. Activated sludge treatment plant in Welland, Ontario (photo courtesy of Mr. P. Laughton).

Figure 9-5. An example of a wooden cover on a secondary clarifier at sewage treatment plant in Gravenhurst, Ontario (photo courtesy of Greey Lightnin).

Figure 9-6. Covered secondary clarifiers in Canada Packers Plant in Fort William, N.S. (photo courtesy of Greey Lightnin).

Energy Conservation. Ever increasing cost of energy over the past decade makes some older operational costs estimates and comparisons for activated sludge treatment not valid any more, and calls for potential saving measures. They are possible above all, by a selection of economical aeration methods, and by an application of an automated dissolved oxygen control (see Chapter V).

Reliability of Power Supply. To avoid emergency discharges of untreated wastewater in the case of a power supply failure, activated sludge plants at some sensitive locations, require proper sources of additional standby power. Usually, they are connected to dual electric services, and/or have an on-site electric power generator. Electric power service reliability varies in different locations, and should be investigated separately for each particular activated sludge treatment plant. The economy of on-site power generation was discussed by Meckler (1976).

Safeguards and Fencing. Similar safeguarding and fencing is required at activated sludge treatment plants as at other types of wastewater treatment plants and water purification plants. However, care has to be taken to provide, when required, the proper guardrails, handrails on steps and gratings to cover pit openings.

Plant Aesthetics and Environmental Considerations. Care should be taken to design activated sludge treatment plant facilities in an aesthetic way, in keeping with the plant surroundings.

Many aeration tanks at activated sludge treatment plants located close
to dwellings are covered to prevent the direct stripping of odorous substances from the mixed liquor into the ambient air. At some plants, grit
chambers and primary tanks are covered also. Exhaust air from below
the covers is deodorized by treatment with ozone, wet scrubbing with
hypochlorite solution, or carbon adsorption. In the past, some odour
masking substances were used to avoid complaints of air pollution by wastewater treatment plants.

Motors and compressors at activated sludge treatment plants can be
a serious source of noise. However, the various types of this equipment
differ substantially in this respect. Therefore, a careful selection of
equipment should be made to suit specific local requirements. In addition,
an appreciable noise reduction can be achieved by proper isolation and
housing of noise sources.

9.6 SMALL PLANTS

Small activated sludge treatment plants, as is the case with large plants
(see p. 9.7) possess somewhat unusual specific design characteristics.
However, most of the information available on the activated sludge process
is generally relevant to the most popular plants which are of medium size.

Small activated sludge treatment plants are usually designed as package (pre-engineered) plants, employing the extended aeration modifications
of the process (see Chapter IV p. 4.2) and operated in a batch or continuous flow mode, with mechanical aeration or aeration by means of diffused
air, and with a not-routine sludge wasting. A typical feature of these
plants is an excessive accumulation of sludge associated with somewhat
elevated contents of suspended solids in the final effluent. The National
Sanitation Foundation (1966) originally formulated criteria for evaluation
of these plants, and this subject was consequently dealt with by several
authors (Environmental Protection Agency, 1977). Figure 9-7 presents
an example of a transportable activated sludge package plant.

Another type of package plants, often suggested for somewhat larger
flows, is based on the contact stabilization modification of the process
(see Chapter IV, p. 4.5). For this type of small activated sludge plant,
the National Sanitation Foundation (1968) also formulated the original
evaluation criteria. Figure 9-8 presents a typical layout of a small to
medium size contact stabilization plant, designed without primary clarification and with aerobic sludge stabilization. Such plants are more sensitive to fluctuating flows and loads, usual for small plants, than the extended aeration plants described before. Dague et al. (1972) presented
two examples of small contact stabilization plants where efficient operation was not possible, and which eventually were converted into other
modifications of the activated sludge process treatment.

Figure 9-7. Transportable activated sludge "Package Plant" (photo courtesy of Greey Lightnin).

9.7 LARGE PLANTS

The substantial capital and operational costs associated with the construction and operation of large activated sludge treatment plants, justify an extremely thorough approach to design parameters, the selection of technological variants, the selection of equipment, etc. Even small percentagewise savings achieved this way can constitute meaningful sums of money. Therefore, extensive studies and experimentation are usually carried out to obtain for accurate and precise information for these plants. For the design of smaller plants this information can be obtained through much simpler procedures.

An example of the above approach is the Emscher Mouth Treatment Plant, West Germany, which was designed for a wastewater flow of almost 460 mgd (20 cu. m/sec). To attain a possibly flawless design for this plant, experiments were made for a number of years in a relatively large pilot installation, accepting flows up to 480 gpm (30 L/sec) for each of its two lines.

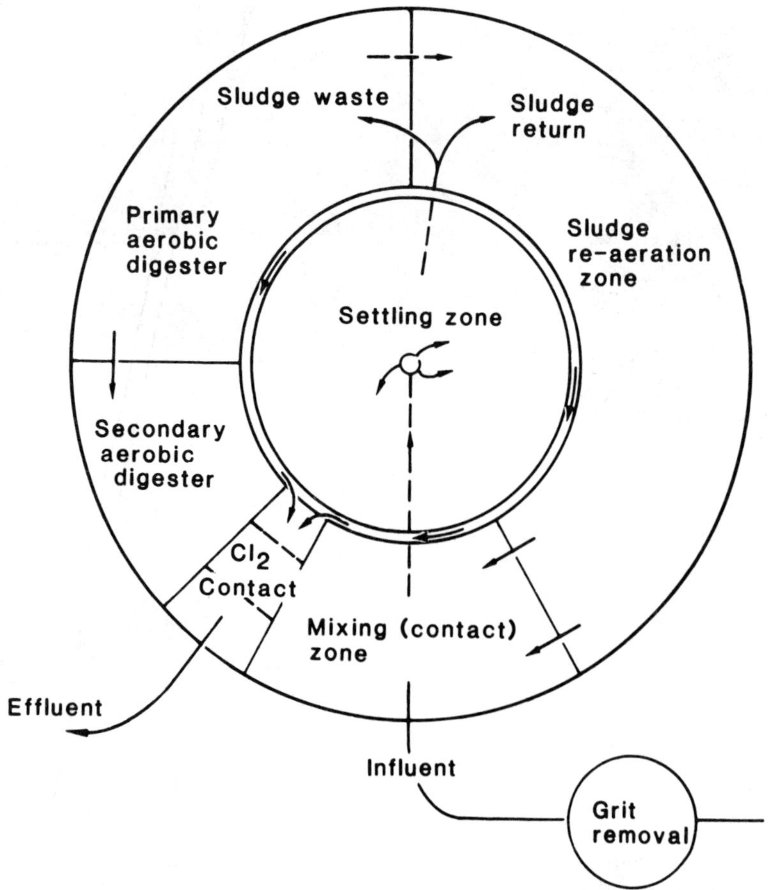

Figure 9-8. Typical layout of a small size contact stabilization plant with an aerobic sludge stabilization.

Most of the large plants for activated sludge process treatment are joint treatment plants, accepting both sewage and industrial effluents. Therefore, the comments on the subject, presented in Chapter VIII, p. 8.3, are also applicable here.

In 1975 and 1979, the International Association on Water Pollution Research organized two Workshops in Vienna, Austria, on the design and operation interactions at large wastewater treatment plants.

REFERENCES

Burkhead, C. E., and McKinney, R. E.: "Application of Complete-Mixing Activated Sludge Design Equations to Industrial Wastes," Jour. Water Poll. Control Fed., 40, 557 (1968).

CIRIA: "Cost-Effective Sewage Treatment - The Creation of an Optimizing Model," CIRIA Report 46, Construction Industry Research and Information Association, London, Great Britain, 1973.

Dague, R. R. et al.: "Contact Stabilization in Small Package Plants," Jour. Water Poll. Control Fed., 44, 255 (1972).

Eckenfelder, W. W., Jr., and O'Connor, D. J.: "The Aerobic Treatment of Organic Wastes," Proc. 9th Ind. Waste Conf., Purdue Univ. Ext. Ser. 89, 2, 215 (1955).

Eckenfelder, W. W., Jr.: "Theory and Practice of Activated Sludge Process Modifications," Water & Sewage Wks. 108, 4, 145 (1961).

Eckenfelder, W. W., Jr.: "Industrial Water Pollution Control," McGraw-Hill, New York, New York, 1966.

Eckenfelder, W. W., Jr.: "Theory of Biological Treatment of Trade Wastes," Jour. Water Poll. Control Fed., 39, 240 (1967).

Eckenfelder, W. W., Jr.: "Water Quality Engineering for Practicing Engineers," Barnes & Noble, New York, New York, 1970.

Environmental Protection Agency: "Process Design Manual: Wastewater Treatment Facilities for Sewered Small Communities," EPA-625/1-77-009, October 1977.

Environmental Protection Agency: "Construction Costs for Municipal Wastewater Treatment Plants: 1973-1977," Office of Water Program Operations, 430/9-77-013, MCD-37, Washington, D.C., January 1978(a).

Environmental Protection Agency: "Analysis of Operations & Maintenance Costs for Municipal Wastewater Treatment Systems," 430/9-77-015, MCD-39, Washington, D.C. May 1978(b).

Floegl, W.: "Vergleichende Kostenuntersuchungen ueber das Beleburgsverfahren," Wiener Mitteilungen, V. 36, Wien, Austria, 1980.

Goodman, B. L., and Englande, A. J., Jr.: "A Unified Model of the Activated Sludge Process," Jour. Water Poll. Control Fed., 46, 312 (1974).

Green, A. J., et al.: "Computer-Assisted Procedure for the Design and Evaluation of Wastewater Treatment System - Users Guide," CE/EPA 430/9-79-10, May 1979.

Laughton, P. J.: personal communications (1980).

Lawrence, A. W., and McCarty, P. C.: "Unified Basis for Biological Treatment Design and Operation," Jour. Sanit. Eng. Div., Proc. Amer. Soc. Civ. Eng., 96, SA3, 757 (1970).

LeClair, B.: personal information (1982).

McKinney, R. E.: "Mathematics of Complete-Mixing Activated Sludge," Jour. San. Eng. Div., Proc. Amer. Soc. Civ. Eng., 88, SA3, 87 (1962).

Meckler, M.: "Preparing an Economic Analysis of an On-Site Energy System," Consulting Engineer, February 1976.

Middleton, A. C., and Lawrence, A. W.: "Least Costs Design of Activated Sludge Systems," Jour. Water Poll. Control Fed., 48, 889 (1976).

Monod, J.: "The Growth of Bacterial Cultures," Annual Review of Microbiology, 3, 371 (1949).

National Sanitation Foundation: "Package Sewage Treatment Plants Criteria Development. Part I - Extended Aeration," September 1966.

National Sanitation Foundation: "Package Sewage Treatment Plants Criteria Development. Part II - Contact Stabilization," June 1968.

Parkin, G. F., and Dague, R. R.: "Optimal Design of Wastewater Treatment Systems by Enumeration," Jour. Sanit. Eng. Div., Proc. Amer. Soc. Civ. Engrs., 98, SA6, 833 (1972).

Patterson, W. L., and Banker, R. F.: "Estimating Costs of Manpower Requirements for Conventional Wastewater Treatment Facilities," Environmental Protection Agency, 17090 DAN 10/71 (1971).

Rossman, L. A.: "Computer-Aided Synthesis of Wastewater Treatment and Sludge Disposal Systems," EPA - 600/2-79-158, Cincinnati, Ohio, December 1979.

Sherrard, J. G., and Lawrence, A. W.: "Design and Operation Model of Activated Sludge," Jour. Envir. Eng. Div., Proc. Amer. Soc. Civ. Eng., 99, EE6, 773 (1973).

Smith, D. W.: "Computer Design of CMAS Systems," Jour. Sanit. Eng. Div., Proc. Am. Soc. Civ. Engrs., 96, SA4, 977 (1970).

Smith, R.: "Preliminary Design and Simulation of Conventional Waste Renovation Systems Using the Digital Computer," Report No. WP-20-9, U.S. Dept. of the Interior, Federal Water Pollution Control Administration, Cincinnati, Ohio, 1968.

Smith, R., Eilers, R. G., and Hall, E. D.: "Executive Digital Computer Program for Preliminary Design of Wastewater Treatment Systems," Report No. WD-20-14, U.S. Dept. of the Interior, Federal Water Pollution Control Administration, Cincinnati, Ohio, 1968.

Stensel, H. D., and Shell, G. L.: "Two Methods of Biological Treatment Design," Jour. Water Poll. Control Fed., 46, 271 (1974).

Tihansky, D. P.: "Historical Development of Water Pollution Control Cost Functions," Jour. Water Poll. Control Fed., 46, 813 (1974).

Operation and Maintenance of Activated Sludge Plants

10.1 GENERAL

Operation and maintenance of an activated sludge treatment plant should
follow the recommendations covered by an "Operation and Maintenance
Manual" prepared specifically for the given plant, and updated periodically
by information gained through experience in the operation of this plant.
General suggestions regarding contents of such manuals were reviewed by
Green et al. (1974). The Water Pollution Control Federation (1976) released a "Manual of Practice No. 11," dealing with the problems of wastewater treatment plant operation. This text can also be used as a basis for
preparation of the individual "Operation and Maintenance Manual."

10.2 STARTING-UP PROCEDURES

Before starting the operation of a new activated sludge treatment plant, a
"dry check-out" of the installation has to be carried out. It consists of a
comparison of the completed plant with information gathered from the design drawings, operation and maintenance manual, and manufacturers'
instructions; lubrication and the dry testing of the equipment; cleaning
tanks and conduits; and, preparation of daily operational logs. Then a
"wet check-out" follows. The facilities are filled with water and the aeration equipment, the mechanical equipment of the secondary clarifier(s),
and the operation of all controls, instrumentation, and other equipment,
are checked. Only after the successful completion of the "wet check-out,"
can the wastewater be slowly directed to the treatment system. Initially,
municipal wastewater should flow only at a fraction of the nominal flow,
and some industrial effluents should be diluted with water, prior to reaching the aeration tank.

The most convenient formulation of the proper activated sludge for the treatment of specific wastewater is by "seeding" the aeration tanks with activated sludge from plants treating wastewater of a similar type. In cases of treatment of so-called "difficult" wastewater, a good strategy is to use, initially, relatively large quantities of sludge, as acclimatization process losses will soon decrease the level of suspended solids in the aeration tanks. However, the initial oxygen uptake by the "seed" sludge load must not surpass the aeration capacity of the aeration system, as the required level of dissolved oxygen has to be maintained throughout the starting-up period and afterwards.

The duration of the starting-up period may differ widely, and depends on a number of factors. The most important of these are the qualifications of operational personnel (see p. 10.7), and in the case of some industrial effluents, the adaptability of the seeding micro-organisms. It is a good strategy to bring in the operators about 3 months before completion of the plant construction, and to allow them to participate in equipment testing, etc. Recently, Johnson (1981) presented an interesting description of the frustrations and accomplishments of the successful start-up of an oxygen activated sludge plant.

10.3 PROCESS CONTROL

CONTROL DATA

The control of activated sludge treatment plants is based on certain flow measurements and the determination of a number of quality factors. The usual control program includes flow measurements of:

> influent wastewater;
>
> recycling and wasted activated sludge;
>
> compressed air to the aeration tanks (if this system of aeration is applied).

Analytical determinations cover:

> influent wastewater quality (BOD = biochemical oxygen demand, COD = chemical oxygen demand, TOC = total organic carbon, total nitrogen, total phosphorus, suspended solids, volatile solids, specific organic and mineral substances, coliform bacteria, etc.); and
>
> effluent wastewater quality (as above, plus the forms of nitrogen: ammonia, nitrates, nitrites, and organic nitrogen);
>
> aeration tanks dissolved oxygen (D.O.) concentrations;
>
> mixed liquor suspended solids, mixed liquor volatile solids, activated sludge volume index, and the recycling and wasted activated sludge suspended solids.

Details of such a control program are given in the respective "Process Control Manual" (Environmental Protection Agency, 1977).

The frequency of measurements and sampling differs in various plants, and often depends on the local possibilities or local regulations. Carr and Ganczarczyk (1971) gave an example of such metering and sampling programs in their paper on the Lakeview Sewage Treatment Plant in Port Credit, Ontario. For this specific case, the meter accuracy was designed with a built-in tolerance of ±2% of the actual flow. One of the problems encountered in the day-to-day operation of the plant was that instrumentation soon lost its initial calibration and hence its accuracy. There were reasons for this: grease present in sewage tended to be deposited on the walls of any measuring device placed in the flow, and coated the lines to the pressure bellows, which were easily clogged. Although most flow metering devices depend on automatic purge water systems to keep lines free, quite often this operation must be performed manually. As this type of maintenance involves much labour and time, plant personnel have a tendency to go by the pump setting alone, regardless of what the meters indicate. Sampling procedures and techniques at Lakeview were later modified by the introduction of automatic samplers. All sampling, whether automatic or manual, was done on a unit basis, usually a 1-liter sample per 8 hours of plant operation (sample size is not proportional to flow). Samples collected are analyzed on a daily basis. Sampling problems include the operator's ability to collect and decant uniform composite samples and to recognize and take a grab sample of any unusual or exotic wastewater. Raw sewage sampling sometimes causes problems, as large particles will often plug up the sampler lines. Weekend samples are often stored in the plant refrigerator for analysis on the following Monday.

In most cases the frequency of sampling in wastewater treatment plants does not take into consideration the plants' characteristic through-flow patterns, e.g., by taking the grab samples of the plant influent and effluent at practically the same time, taking the grab samples at the same hour of the day, etc. Therefore, a definite bias occurs in the results obtained. However, this is not a problem when information is based on daily composite samples. Also, a certain degree of variation must be expected when flow measurement and chemical analysis for treatment control are conducted. If, in the latter case, the average of several determinations is taken as the true determination, it must be remembered that this average is a measurement obtained from one sample. Therefore, the subject of sampling error has to be considered. The validity of the results obtained depends on the fairness of the sample and the technique employed in studying that sample. To secure representative control data, it is necessary to obtain characteristic samples selected on the basis of the previously known variations of the treatment performance. Of course, they need not be simple random samples, but may be properly stratified samples anticipating the known fluctuation patterns. Therefore, to choose appropriate sampling procedures, extensive study of the plant is needed. It also should be mentioned that the accuracy of most analytical determinations used in the control of the treatment processes is not known clearly enough for application to particular cases. It seems that there is a serious need for an examination of this question.

As was stated earlier, the only way to prevent systematic errors in treatment plant control data is to collect the data according to a precise program, based on knowledge of the process and the plant. It is also possible to increase, if necessary, the accuracy of most of the analytical procedures used, by selecting more advanced analytical methods and using more precise instruments. The only question is how many data are really needed for the practical task of operating the given plant with the required performance efficiency, and how accurate the data must be. This question is by no means an easy one, but it has to be answered if efficient and economical operation of the plant is to be achieved.

If, in a given treatment plant, there is a shotage of manpower, there is a possibility that the analytical data received will not be sufficient and the plant operation will be heavily based on the operator's intuition. On the other hand, if there is a surplus of manpower or automatic analytical equipment is available, the flood of control data causes the additional and difficult task of data storage and retrieval. There is a feeling that in some cases extensive and frequent measurements and analysis for the control of treatment plants have only very limited practical significance and are often used only to fill the monthly and annual reports. At the same time, for immediate plant control, only limited measurements are made. Such a simplification, however, should not be abused and its use is justified only if the particular plant has been previously studied to a reasonable extent and no surprise effects are expected.

ANALYSIS OF CONTROL DATA

Using different statistical techniques, wastewater treatment plant performance data can be analyzed and evaluated. The most common approaches are frequency analysis, time-series analysis, and correlation of technological parameters. The basic frequency analysis includes calculations of mean values, standard deviations, variances, and coefficients of variation. Usually the influent BOD shows the largest standard deviation from the mean, while variations become progressively less as the wastewater is given first physical (primary) and then biological (secondary) treatment. However, the coefficients of variation of secondary effluent BOD and suspended solids are often substantially higher than for primary effluent. This may be explained by the lower stability of the biological treatment than of the mechanical treatment, due to the complexity and sensitivity to environmental changes of the former. The changes in the coefficient of variation with increasing treatment may also be regarded as a measure of the plant's buffering capacity or ability to accommodate the qualitative and quantitative shock loads imposed on it. Therefore, an increase in coefficient of variation may indicate insufficient buffering capacity in the system studied. However, the data on final effluents may also reflect substantial "background noise" representing inadequacies in the analytical equipment at this relatively low range of concentration.

It is also interesting to note that most data on wastewater quality show the log-normal distribution or some other similar skewed distribution. This fact has been confirmed in many studies of wastewater treatment and may be used for a more precise description of studied or projected situations where, often, normal probability density functions are assumed or considered to exist. It has proved convenient to present some complicated relationships in the wastewater treatment processes in the form of comparisons of frequency distributions, rather than by using regression analysis or deterministic mathematical models. This is especially true of phenomena whose response time is difficult to determine and/or fluctuating during the observation period. It is possible using time-series analysis to determine quantitatively all significant trends and periodicities in wastewater treatment plant performance data. Seasonal or any special kind of variability can be removed from the data as given harmonic components in a Fourier analysis. The residual record can be the subject of autocorrelation and spectrum analysis.

There are, however, specific difficulties in "data editing" and "trend correcting" when the effects of a biological treatment are being studied. It seems that, sometimes, quite substantial differences may be expected as the result of responses of some systems studied to particular external impulses, so rejection of some of the measured data in the process of editing is not necessarily justified if not supported by additional studies.

The study of the relationships among various technological parameters of treatment methods can be used to prove the correctness of the applied metering and sampling systems, and by itself supplies information vital for the most efficient operation of the plant. There are, however, specific difficulties in the realization of this approach.

Most of our knowledge of the treatment processes is based on experiments carried out under so-called "controlled laboratory conditions." Such studies can be much broader and more diversified than experiments in pilot or full-scale operations, but are often biased by the undue influence of some factors which cannot be properly controlled or modelled. It is usually taken for granted that the behaviour of the full-scale plants will be close to that of the models studied, and the relationships among process parameters, observed in smaller scale, will be equally valid in the full-scale plants. However, quite often the fluctuation of wastewater quantity and quality, as well as the instability of some process parameters, together with the inadequacies of metering and sampling, are such that in specific cases expected correlations cannot be realized in practice. The data are characterized by such a scatter of information that their correlation would be without any practical meaning.

CONTROL STRATEGIES

The strategies for control of the activated sludge process are based on
maintenance of the required:

1. mixed liquor suspended solids level;

2. F/M (food to micro-organisms) ratio, which comprises concepts of
    so-called "load balancing" (the concept of respiration rate control
    could be considered to be a derivative of this approach); and

3. sludge age (or mean cell residence time—MCRT) control.

Generally, three solids inventory control modes are possible in an
activated sludge system. These include:

1. simple control of the recycle flow rate;

2. control of the recycle flow rate when provision has been made for a
    constant volume storage chamber; and

3. control of the recycle flow rate when provision has been made for
    a variable volume storage chamber.

The first strategy is not a suitable means for controlling the large diurnal
flow variations experienced in many treatment plants. However, all the
above methods could be used both for the F/M ratio control and MCRT
control.

Another approach to the F/M ratio control may be based on frequent
measurements of the mass rate of flow of biodegradable organic substances
into the aeration tanks, divided by the mass rate of flow of return sludge
(Kugelman et al., 1977). Attempts have been made to apply TOC to measure both F and M. In Philadelphia's North-East Water Pollution Control
Plant (DiMenna and Wankoff, 1977), waste activated sludge, which is
pumped to the primary sedimentation tanks, is withdrawn from the system
on a scheduled basis to avoid adding any additional load to the plant at peak
loading periods, and to help equalize the instantaneous food to micro-
organism ratio throughout the day. Operation of the activated sludge system is modified in winter months to reflect decreased biological activity at
lower temperatures. Wasting of activated sludge directly from aeration
tanks offers many operational advantages (Burchett and Tchobanoglous,
1974).

Control of the activated sludge process by respirometric measurements
of mixed liquor under actual loading condition (Benefield et al., 1975;
Arthur, 1976; and Brouzes, 1979) can be accomplished by the application
of various on-line instruments. Generally, two different types of information may be produced this way: an analogue graph of the actual dissolved
oxygen values during the measurement time, and a trend chart of the

maximum dissolved oxygen value which occurs at the end of each cycle. Both graphs provide valuable information about the wastewater tested.

Cashion et al. (1979) analyzed different activated sludge process control strategies, and found that residual suspended solids accounted for a major portion of the total carbonaceous material present in the plant effluent. Reductions in soluble organic matter that were obtained by solids inventory control, were offset by those residual suspended solids. Therefore the F/M ratio control gave no overall net benefits under the conditions studied.

AUTOMATED CONTROL

Activated sludge treatment units are difficult to control automatically. Some of the reasons for this are the complex process kinetics and the relatively long detention times, often connected with a complicated mixing pattern. Because of the latter, the feedback control systems (quality sensors located in the effluent) often work erratically. More promising is feed forward control (sensors in the influent).

There have been several theoretical and practical attempts to control automatically the activated sludge process. In general, to design a workable automatic control system, reliable and accurate methods of automatic measurement and monitoring of the following parameters are required:

1. flow of wastewater and recycling activated sludge;

2. dissolved oxygen and mixed liquor suspended solids concentrations in aeration tanks; and

3. BOD of influent and effluent.

At present, a satisfactory technique has been developed only for the flow measurements (Water Pollution Control Federation, 1978) and the dissolved oxygen concentration (see Chapter V, p. 5.8). At first, the redox potential measurements were suggested to automate the activated sludge process operation. These attempts, however, were not successful. Also, the automatic measurement of mixed liquor suspended solids concentrations and wastewater BOD is not yet reliable enough, although diverse technical equipment is offered for this purpose. Its application may be connected with some operational problems.

Most existing methods for the automatic determination of mixed liquor suspended solids concentrations in aeration tanks depend on the measurement of the transmission or the scattering of visible light, or some other source of radiation, as in automatic determinations of turbidity. These optical properties of mixed liquor must be related to the actual suspended solids contents of a specific activated sludge. This leads to significant errors, which are due to the substantial concentration and fluctuating

flocculation abilities of the material, as well as its other specific features,
e.g., distribution of floc sizes. In the condition of most aeration tanks,
this distribution may be bimodal, making this material completely unsuit-
able for any optical concentration measurements. Some correction of these
measurements is possible by compensation for mixed liquor soluble colour
through simultaneous measurements of optical density at two different
wavelengths, or simultaneous measuring of scattered and transmitted light.
These improvements may have some applications in the monitoring of suspended solids in river water or treated effluents, but, in mixed liquor,
their role is very limited.

Different attempts to measure BOD automatically have not as yet produced the required information. Also, substituting COD determinations or
organic carbon measurements for BOD measurements does not seem satisfactory. In recent years extensive studies have been carried out on the
possible application of some enzymatic tests for automation of the activated
sludge process. A review, in Japan, of available automatic wastewater
quality analyzers was recently presented by Murakami (1980).

Several automated pilot facilities for activated sludge treatment were
described. Klei and Lundstrom (1974) described an automated control system for a pilot activated sludge reactor. This system proved capable of
regulating air flow and activated sludge recycle to compensate for the
fluctuating carbon concentration in the reactor influent. The feed stream to
the reactor after automatic centrifugation was automatically analyzed for
total carbon concentration every four minutes. The signals from the infrared analyzer were sent to a recorder equipped with a retransmitting slide
wire, and followed by a peak follower, coupled with a converter. In another
pilot-plant study, Stephenson et al. (1981) investigated the use of continuous
on-line sensors (for flows, dissolved oxygen and suspended solids) and a
real-time computer for automated data acquisition and control of the activated
sludge process. Dynamic step feed control based on specific oxygen uptake
rate, dissolved oxygen control and solids inventory control through solids
retention time and volumetric oxygen uptake rate control were studied.
The feasibility of the above automated control strategies as proven, and the
treatment results achieved were equal to those of a well-controlled, man-
ually-operated plant.

LABORATORY CONTROL

In addition to the previously described analytical control tasks, laboratories
at activated sludge treatment plants, can be involved in various studies to
optimize the system operated and to overcome some operational problems,
etc. The respective control procedures were described in detail by West
(1978). They include also several unconventional approaches, which although questionable from a theoretical point of view, like a centrifuge
method for determination of mixed liquor suspended solids concentrations
(Setter, 1935; and Drews, 1978), were found practical under plant

conditions. Recently, more and more of a laboratory tests are becoming automated or semi-automated.

ALARM SYSTEMS

Although modern activated sludge plants are usually designed to operate unattended, operational accidents may cause interruption of service and/or equipment damage. To prevent such situations, proper alarm systems have to be installed to signal for operational personnel assistance when any malfunction occurs in the vital elements of the plant.

During the working hours of the operational personnel, an alarm in the form of a light or sound signal at the plant master control panel (Fig. 10-1) is satisfactory. However, for the after-hours alarm, it is necessary to apply a more sophisticated system, connected to the home telephone numbers of the selected operational personnel. This system is tied to a reporting set, connected with the master control panel and programmed to convey the information about the abnormal conditions of the operation by means of a selected pre-recorded magnetic tape (Fig. 10-2). It is possible to program the alarm system to dial the home telephone numbers of the

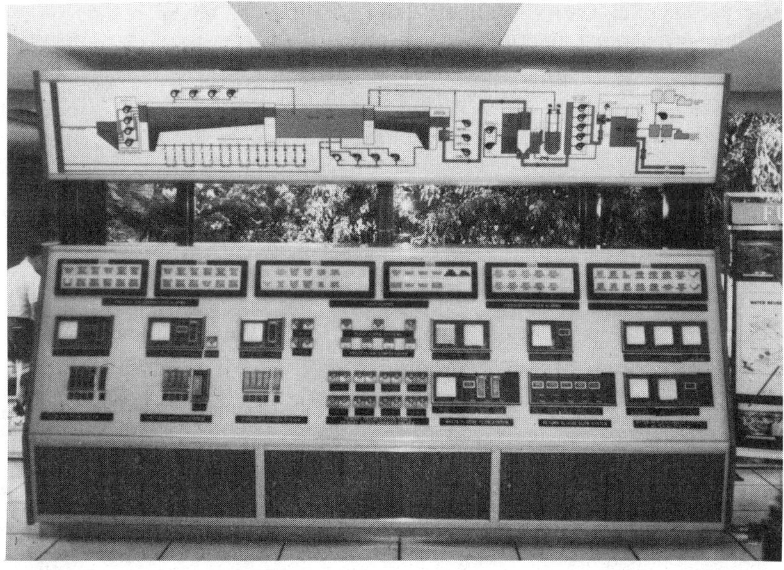

Figure 10-1. Control panel of an activated sludge process treatment plant (photo courtesy of P. Laughton).

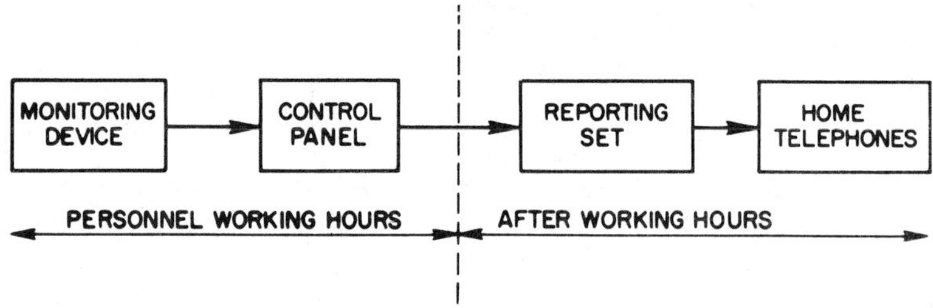

Figure 10-2. Alarm systems for an activated sludge process treatment plant.

operational personnel in sequence, according to their likely availability, or to use of a computerized telephone call diverter operating in accordance with a selected sequence.

The alarm could be activated by power failure, phase failure, unusually high or low flows of wastewater, recycle and air, low dissolved oxygen levels, and low suspended solids in the mixed liquor, etc. It is a wise strategy to program the home alarm system to start only if the above factors are reaching levels seriously affecting the plant operation. Minor malfunctions should only be automatically recorded and subsequently corrected during the working hours of operational personnel.

10.4 COMPUTER FACILITATED OPERATION

At present, it appears that the application of a computer assisted plant operation may be more successful than some premature attempts to broadly apply automation which was not developed well enough. Specifically, computers are being used mostly for:

control of all flows of wastewater, sludges and air, including flow splitting operations, etc.;

operational data processing, monitoring, storage and retrieval;

automation of aeration systems.

Usually, computer capacities are much larger than needed for the above tasks. The spare computer capacity is left for future needs which may arise with the development of new sensors. The same computer can be also used for other tasks such as the control of a combined sewer system, etc.

Computers also can accept various process control models and use them for operational advice formulation, or even for automated control. In an example of the computer assisted operation of an activated sludge plant (Trussel et al., 1974), the computer catalogues plant operating and laboratory data, conducts a mass balance on the plant and uses data from the past five days and the principles of activated sludge kinetics to develop recommendations which are designed to maintain a constant sludge retention time (sludge age). Provision is also made for periodic adjustment of the sludge load (the food to micro-organisms ratio) in the event that the control based on the sludge retention time does not respond to changes in loading with sufficient speed. Von Jeszenszky and Dunn (1976) developed a dynamic control model for a two-stage (single sludge or two sludges) activated sludge treatment with feedback or feedforward control possibilities. In a computer simulation study it was found that feedback on-off control was satisfactory for a linear growth approach, whereas feedforward control was more effective for the non-linear Monod kinetics. Nelson and Mishra (1980) described a digital on-line closed-loop control system used in the Denver oxygen activated sludge plant. The system includes digital feedback control of influent flow-splitting, return sludge flow control using a food to mass ratio, waste activated sludge mass control, process quality parameter monitoring, etc.

10.5 OPERATIONAL PROBLEMS

Generally, the operational problems of activated sludge treatment plants can be divided into: mechanical problems (connected with malfunctions of some pieces of equipment) and technological problems (causing a decrease in final effluent quality). The former problems usually are solved by proper repairs or replacements, but solutions to the latter ones may not be so simple. The technological problems mostly arise from the inadequate separation of activated sludge from the treated effluents (see Chapter II and VI) but may also be associated with some decreases in substrate removal rates (see Chapter II). The sludge separation problems may be classified as: sludge bulking, rising of sludge, formation of dispersed growth and pin floc, and formation of foam and scum.

<u>Sludge bulking</u>. The mechanisms of sludge bulking and the respective remedies are described in Chapter III.

<u>Rising of sludge</u>. Rising of sludge to the surface of secondary clarifiers is usually caused by sludge flotation due to the molecular nitrogen development through denitrification of nitrates. Therefore, this phenomenon is limited to nitrifying sludges, and can be avoided or minimized by an increase in the recirculation rate, which increases the level of dissolved oxygen in the contents of the clarifiers.

<u>Formation of dispersed sludge and pin-point sludge</u>. Formation of poorly

flocculating, dispersed sludge is a function of too high sludge loadings
(see Chapter III). A decrease in the F/M ratio should decrease the residual
level of suspended solids in the final effluent. Pin-point flocs are characterized by the absence (or a serious limitation) of filamentous growths in
the sludge. This may be due to some industrial components of the treated
wastewater.

<u>Formation of foam and scum.</u> Formation of foam in aeration tanks may be
a consequence of the presence in wastewater of foaming agents such as
some industrial chemicals, by-products of protein decomposition, etc.
Heavy metals in wastewater may stabilize the formed foam (Laughton,
1981), and aeration of excess activated sludge in the process of aerobic
digestion may be connected with extensive foaming. Another kind of a
heavy foam and scum formation may be caused by growth of some actino-
mycetes like Nocardia app. (Pipes, 1978, and Dhaliwal, 1979). It appears
that the growth of these organisms is associated with the presence of
emulsified oil and grease in wastewater.

   A remedy for foam formation is a mechanical defoaming by application
of sprays (Fig. 10-3), vacuum fans, special turbines (Ng and Gutierrez,
1977), increase in mixed liquor suspended solids levels, or application of
some anti-foaming additives (higher alcohols, silicon oils, etc., which are
effective at a dosage in the range of 1 mg/L).

Figure 10-3. Foam control by application of sprays (photo by the author).

<u>Other problems</u>. Operation of industrial activated sludge plants during production breaks creates a difficult problem which can be approached only on the basis of specific local conditions.

10.6 PREVENTIVE MAINTENANCE

Preventive maintenance of an activated sludge plant (Fig. 10-4) protects the investment by increasing the life of the structures and the equipment, and it is, therefore, a profitable operation which also improves plant operation reliability. The maintenance schedule should be worked out on the basis of the recommendations of the equipment suppliers and the plant design consultants. Proper records of the maintenance steps should be kept. The maintenance schedule can be prepared in the form of charts, a card file, or, if the plant is equipped with a computer, it may be programmed to produce printouts calling for proper maintenance steps on specified days.

At smaller activated sludge plants, the maintenance responsibilities are often simplified by contracting out at least part of the required work

Figure 10-4. Diffuser tubes removed for inspection or cleaning (photo courtesy of P. Laughton).

to the equipment suppliers. Such an approach has some merits, as equipment suppliers often have highly qualified maintenance technicians on their
staff.

10.7 OPERATIONAL PERSONNEL

One of the weakest points in the operation and maintenance of activated
sludge treatment plants is the quality of operational personnel. It was
shown in several examples that many plants performed poorly only because the operating staff did not understand the treatment process principles
and did not have the technical skills to properly operate, control and maintain the treatment facilities. Quite often, however, these operators were
also receiving insufficient or inaccurate instructions and recommendations.

As a rule, large treatment plants have more competent and better
trained staff; smaller plants often are not financially able to employ people
of comparable caliber. In industry, activated sludge treatment plant
operators also often have other responsibilities and, if they prove capable,
are sometimes shifted to other "more important" duties.

To upgrade the skills of the operators, various training courses are
being organized either in special training centres or through technical
assistance on the job site. The former type of training is often connected
with some type of certification.

The staffing needs for activated sludge process plants are estimated
on the basis of various guidelines, and can differ widely, An interesting
analysis of the subject is presented by Culp et al. (1977).

REFERENCES

Arthur, R. M.: "The On Line Respirometer and Its Use in Operation Control to Save Energy and Reduce Cost," paper presented at the 49th WPCF
Conference, Minneapolis, Minnesota, 1976.

Benefield, L. D. et al.: "Process Control by Oxygen-Uptake and Solids
Analysis," Jour. Water Poll. Control Fed., 47, 2498 (1975).

Brouzes, P. H.: "Monitoring and Controlling Activated Sludge Plants in
Connection with Aeration," Prog. Wat. Tech., 11, 3, 193 (1979).

Burchett, M. E., and Tchobanoglous, G.: "Facilities for Controlling the
Activated Sludge Process by Mean Cell Residence Time," Jour. Water
Poll. Control Fed., 46, 973 (1974).

Carr, D. F., and Ganczarczyk, J.: "A Performance Analysis of an Activated Sludge Treatment Plant," Publ. No. 71-6, University of Toronto,
Dept. of Civil Engin., Toronto (1971).

Cashion, B. S. et al.: "Control Strategies for Activated Sludge Process," Jour. Water Poll. Control Fed., 51, 815 (1979).

Culp, G. et al.: "Evaluating Wastewater Facility Staffing Needs," Jour. Water Poll. Control Fed., 49, 2226 (1977).

Dhaliwal, B. S.: "Norcardia amarae and Activated Sludge Foaming," Jour. Water Poll. Control Fed., 51, 332 (1979).

DiMenna, R. A., and Wankoff, W.: "Operational Changes Improve Treatment at Philadelphia's NE WPT Plant," paper presented at the 50th WPCF Conference, Philadelphia, Pennsylvania, 1977.

Drews, R. J. L. C.: "A Rapid Centrifuge Method for Determination and Control of Sludge Concentration in Activated Sludge Plants," Water SA, 4, 1, 1 (1978).

Environmental Protection Agency: "Process Control Manual for Aerobic Biological Wastewater Treatment Facilities," EPA-430/9-77-006, Municipal Operations Branch, March 1977.

Garret, M. T., Jr.: "Hydraulic Control of Activated Sludge Growth Rate," Sew. and Ind. Wastes, 30, 253 (1959).

Gilbert, W. G.: "Municipal Wastewater Treatment - The Relation of Operation and Maintenance Activities to Plant Efficiency," paper presented at the 48th Annual Conference of the Water Pollution Control Federation, Miami Beach, Florida, 1975.

Green, R. L., et al.: "Considerations for Preparation of Operation and Maintenance Manuals," EPA 430/9-74-001, Washington, D.C., 1974.

Von Jeszenszky, T., and Dunn, I. J.: "Dynamic Modelling and Control Simulation of a Biological Wastewater Treatment Process," Water Research (Brit.), 10, 461 (1976).

Johnson, J. A.: "Start-up and Operation of an Oxygen-Activated Advanced Wastewater Treatment System," Jour. Water Poll. Control Fed., 53, 451 (1981).

Klei, H. E., and Sundstrom, D. W.: "Automatic Control of an Activated Sludge Reactor," Jour. Water Poll. Control Fed., 46, 993 (1974).

Kugelman, I. J. et al.: "Activated Sludge Control with the Aid of Instantaneous F/M, Aerator Respiration Rate and Settleometer Measurements," paper presented at the 50th WPCF Conference, Philadelphia, Pennsylvania, 1977.

Laughton, P. J.: personal information (1981).

Murakami, K.: "Automatic Water Quality Analyzers for Wastewater Collection and Treatment," Jour. Water Poll. Control Fed., 52, 938 (1980).

Nelson, J. K., and Mishra, B. B.: "Digital On-Line Closed-Loop Control
for Wastewater Treatment Operation," Jour. Water Poll. Control Fed.,
52, 406 (1980).

Ng, K. S., and Gutierrez, L.: "Mechanical Foam Breaker-Means for Foam
Control in Wastewater Treatment," Jour. Water Poll. Control Fed., 49,
2310 (1977).

Pipes, W. O.: "Actinomycete Scum Production in Activated Sludge Process," Jour. Water Poll. Control Fed., 50, 628 (1978).

Setter, L. R.: "A Rapid Determination of Suspended Solids in Activated
Sludge by the Centrifuge Method," Sewage Wks. Jour., 7, 23 (1935).

Srinivasaraghavan, R., and Gaudy, A. F., Jr.: "Operational Performance
of an Activated Sludge Process with Constant Sludge Feedback," Jour.
Water Poll. Control Fed., 47, 1947 (1975).

Stephenson, J. P. et al.: "Automatic Control of Solids Retention Time and
Dissolved Oxygen in the Activated Sludge Process," paper presented at
the Internat. Workshop on Practical Experience of Control and Automation
in Wastewater Treatment, Munich and Rome 1981.

Trussell, R. R. et al.: "Computer Assisted Operation of an Activated
Sludge Plant," paper presented at the 47th Annual Conference of the Water
Pollution Control Federation, Denver, Colorado, 1974.

Water Pollution Control Federation: "Operation of Wastewater Treatment
Plants," Manual of Practice No. 11, Washington, D.C., 1976.

Water Pollution Control Federation: "Instrumentation in Wastewater Treatment Plants," Manual of Practice No. 21, Washington, D.C., 1978.

West, A. W.: "Operational Control Procedures for the Activated Sludge
Process," EPA-430/1-78-016, National Training and Operational Technology Center, Cincinnati, Ohio, 1978.

Glossary

Activated Sludge Process—A method of biological wastewater treatment involving aeration with flocculating microbial biomass.

Aeration—Introduction of oxygen into activated sludge systems.

Aeration Tanks—Reactors for activated sludge process in which the aeration takes place.

Aeration Time—Retention of wastewater in aeration tanks.

Aerators—Equipment for achieving aeration.

Aerobic Treatment—Biological treatment processes requiring a presence and an uptake of oxygen.

Batch Reactors—Reactors for non-continuous treatment (fill-and-draw reactors).

Bio-degradability—An ability of some organic substances to be biologically decomposed (see also Induced Bio-Degradability).

Biosorption—A modification of the activated sludge process characterized by a relatively short contact time of micro-organisms with the treated wastewater, followed by a "re-aeration" of the biomass (this process modification is also called Contact Stabilization).

Contact Stabilization—(see Biosorption).

Conventional Modification of the Activated Sludge Process—A modification of the process which is characterized by relatively average values of the treatment parameters. The features of this modification are often used as a reference in comparison with characteristics of other modifications.

Deep-Shaft Activated Sludge Process—A modification of a highly efficient
activated sludge treatment performed in special type aeration reactors
(deep shafts).

Dry Check-Out—An initial step in starting-up an operation of activated
sludge treatment facilities.

Excess Sludge—An increase in mass and volume of activated sludge during
an activated sludge process treatement (called also Waste Activated Sludge).

Extended Aeration—A modification of activated sludge treatment characterized by a relatively long retention of wastewater in aeration tanks.

Fill-and-Draw Biological Reactors—(see Batch Reactors).

Fixed Activated Sludge Process—Retention of a part of biomass in aeration
reactors.

Fluidized-Bed Activated Sludge Process—A modification of the activated
sludge process performed in fluidized-bed reactors which are characterized by a very high accumulation of the biomass.

High-Efficiency Modifications of Activated Sludge Process—Modifications
of the process characterized by high aeration capacities of the reactors and
high levels of available micro-organisms in these reactors. The resulting
treatment effects are close to those of conventional treatment but are
achieved at much shorter hydraulic retention times.

Hydraulic Control of Activated Sludge Process—Wasting activated sludge as
mixed liquor removed directly from aeration tanks.

Hydraulic Detention Time—Period of wastewater aeration with activated
sludge.

Induced Biodegradability—Improving biodegradability of some wastewater
organic components by a chemical pretreatment such as oxidation or
reduction.

Joint Treatment—Treatment of effluents from industrial plants with municipal
wastewater, or treatment of wastewater from more than one municipality in
one plant.

Mean Cell Residence Time—Is called commonly the sludge age, and is
equal to the ratio between sludge inventory in an activated sludge system or
in aeration tanks only, and the daily sludge wastage.

Mixed Liquor—The contents of aeration tanks being a mixture of wastewater and recycling activated sludge.

Mixed Liquor Suspended Solids—The concentration of suspended solids in mixed liquor.

Modified Aeration—A modification of an activated sludge treatment, characterized by a relatively short time aeration of wastewater with relatively small amounts of activated sludge. Only a partial treatment of wastewater is achieved this way.

Oxygen Activated Sludge Process—Modifications of the process in which commercial oxygen is used for microbial respiration.

Phostrip Process—An effective removal of phosphorus from wastewater by an application of a special modification of the activated sludge process ("Phostrip" is a tradename).

Pre-Treatment—Application of various treatment steps prior to the discussed wastewater treatment.

Sludge Age—(see Mean Cell Residence Time).

Sludge Bulking—A substantial increase in the sludge volume index values.

Sludge Volume Index—A volume in mL of 1 g of the activated sludge, after a 1/2 hr settling of the mixed liquor.

Sludge Loading—A ratio between the daily load of wastewater BOD and the sludge inventory in the aeration tanks.

Solids Residence Time—A synonym for the mean cell residence time.

Step-Loading Modification of the Activated Sludge Process (called also the step-aeration modification)—Is based on introducing of wastewater to the aeration tanks in a number of points along the line of the mixed liquor flow.

Thermophilic Activated Sludge—An operation of the activated sludge process at constant temperatures close to 122°F or 140°F (50°C or 60°C, respectively).

UNOX Process—A most common application of the oxygen activated sludge processes ("UNOX" is a tradename).

Volumetric Loading—A ratio between the daily load of wastewater BOD and the sludge inventory in the aeration tank.

<u>Waste Sludge</u>—(see Excess Sludge).

<u>Wet Check-Out</u>—A second step in starting-up an operation of activated sludge treatment facilities.

Conversion Tables

A. From English Measures to Metric

Multiply	By	To Obtain
Btu	251.98	cal
cu. ft	28,317	cu. cm
cu. ft	0.0283	cu. m
cu. ft per lb	0.0623	cu. m per kg
cu. ft per min	28.32	L per min
cu. in	0.0164	L
ft	0.3048	m
ft per min	30.48	cm per min
g (US)	3.785	L
g (US) per min	3.785	L per min
HP (horse power)	0.7457	kW (kilo-Watt)
in	2.54	cm
lb	0.4536	kg
lb per 1,000 cu. ft	16.02	g per cu. m
lb per sq. ft	4.8824	kg per sq. m
sq. ft	0.0929	sq. m
sq. in	6.45	sq. cm

B. From Metric Measures to English

Multiply	By	To Obtain
cal.	3.97×10^{-3}	Btu
cm	0.3937	in
cm per min	0.3937	ft per min
cu. cm	3.52×10^{-5}	cu. ft
cu. m	35.3	cu. ft
cu. m per kg	16.05	cu. ft per lb
g per cu. m	0.0624	lb per 1,000 cu. ft
kg	2.205	lb
kg per sq. m	0.2048	lb per sq. ft
kW (kilo-Watt)	1.341	HP (horse power)
L	0.035	cu. ft
L	0.264	g (US)
L per min	0.126	cu. ft per min
L per min	0.264	g (US) per min
m	3.281	ft
sq. cm	0.115	sq. in
sq. m	10.76	sq. ft

C. Temperature Conversion from °F to °C
$$°C = 5/9 \ (°F - 32)$$

°F	°C	°F	°C	°F	°C
32	0	55	12.8	75	23.9
40	4.4	60	15.6	80	26.7
45	7.2	65	18.3	85	29.4
50	10.0	70	21.1	90	32.2

Index